Springer Theses

Recognizing Outstanding Ph.D. Research

For further volumes:
http://www.springer.com/series/8790

Aims and Scope

The series "Springer Theses" brings together a selection of the very best Ph.D. theses from around the world and across the physical sciences. Nominated and endorsed by two recognized specialists, each published volume has been selected for its scientific excellence and the high impact of its contents for the pertinent field of research. For greater accessibility to non-specialists, the published versions include an extended introduction, as well as a foreword by the student's supervisor explaining the special relevance of the work for the field. As a whole, the series will provide a valuable resource both for newcomers to the research fields described, and for other scientists seeking detailed background information on special questions. Finally, it provides an accredited documentation of the valuable contributions made by today's younger generation of scientists.

Theses are accepted into the series by invited nominated only and must fulfill all of the following criteria

- They must be written in good English
- The topic of should fall within the confines of Chemistry, Physics and related interdisciplinary fields such as Materials, Nanoscience, Chemical Engineering, Complex Systems and Biophysics.
- The work reported in the thesis must represent a significant scientific advance.
- If the thesis includes previously published material, permission to reproduce this must be gained from the respective copyright holder.
- They must have been examined and passed during the 12 months prior to nomination.
- Each thesis should include a foreword by the supervisor outlining the significance of its content.
- The theses should have a clearly defined structure including an introduction accessible to scientists not expert in that particular field.

Lauro Oliver Paz Borbón

Computational Studies of Transition Metal Nanoalloys

Doctoral Thesis accepted by
University of Birmingham, United Kingdom

 Springer

Author
Dr. Lauro Oliver Paz Borbón
Theory Department
Fritz-Haber-Institut der Max-Planck
  Gesellschaft (FHI)
Faradayweg 4-6
14195 Berlin
Germany
e-mail: borbon@fhi-berlin.mpg.de

Supervisor
Prof. Roy L. Johnston
School of Chemistry
University of Birmingham
Edgbaston
Birmingham, B15 2TT
UK
e-mail: r.l.johnston@bham.ac.uk

ISSN 2190-5053   e-ISSN 2190-5061

ISBN 978-3-642-18011-8   e-ISBN 978-3-642-18012-5

DOI 10.1007/978-3-642-18012-5

Springer Heidelberg Dordrecht London New York

© Springer-Verlag Berlin Heidelberg 2011

Cover design: eStudio Calamar, Berlin/Figueres

Printed on acid-free paper

Springer is part of Springer Science+Business Media (www.springer.com)

Dedicada a mi abuelo Salvador, por todo lo que he aprendido de él, además de ser siempre, un gran ejemplo a seguir en mi vida

****

I dedicate this thesis to my grandfather, Salvador, for all the things I have learned from him throughout my life, as well as to be an exceptional role model to follow...

Parts of this Thesis have been Published in the Following Journal Articles

Paz-Borbón, L. O.; Johnston, R. L.; Barcaro, G.; Fortunelli, A.; *A Mixed Structural Motif in 34-atom Pd-Pt Clusters*, J. Phys. Chem. C 2007, **111**, 2936–2941.
*Reproduced with permission.

Paz-Borbón, L. O.; Mortimer-Jones, T. V; Johnston, R. L.; Posada-Amarillas, A.; Barcaro, G.; Fortunelli, A.; *Structures and Energetics of 98-atom Pd-Pt Nanoalloys: Potential Stability of the Leary Tetrahedron for Bimetallic Nanoparticles*, Phys. Chem. Chem. Phys. 2007, **9**, 5202–5208.
*Reproduced with permission.

Paz-Borbón, L. O.; Johnston, R. L.; Barcaro, G.; Fortunelli, A.; *Structural Motifs, Mixing and Segregation Effects in 38-atom Binary Clusters*, J. Chem. Phys. 2008, **128**, 134517.
*Reproduced with permission.

Paz-Borbón, L. O.; Gupta, A.; Johnston, R. L.; *Dependence of the Structures and Chemical Ordering of Pd-Pt Nanoalloys on Potential Parameters J. Mater. Chem.* 2008, **18**, 4154–4164.
*Reproduced with permission.

Pittaway, F.; Paz-Borbón, L. O.; Johnston, R. L.; Arslan, H.; Ferrando, R.; Mottet, C.; Barcaro, G.; Fortunelli, A.; *Theoretical Studies of Palladium-Gold Nanoclusters: Pd-Au Clusters up to 50 atoms J. Phys. Chem. C* 2009, **113**, 9141–9152.
*Reproduced with permission.

Paz-Borbón, L. O.; Johnston, R. L.; Barcaro, G.; Fortunelli, A.; *Chemisorption of CO and H on Pd, Pt and Au Nanoclusters Eur. Phys. J. D* 2009, **52**, 131–134.
*Reproduced with permission.

I have also contributed to the following publication:

Logsdail, A.; Paz-Borbón, L. O.; Johnston, R. L.; *Structures and Stabilities of Platinum-Gold Nanoclusters J. Comp. Theo. Nanosci.* 2009, **6**, 857–866.

Supervisor's Foreword

The research described in Oliver Paz-Borbón's Ph.D. Thesis involves the combination of (Gupta-type) empirical many-body potential energy functions and Density Functional Theory (DFT) calculations to study the structures, bonding and chemical ordering of metallic and bimetallic "nanoalloy" clusters and to investigate the chemisorption of hydrogen and carbon monoxide (CO) on bimetallic clusters. Much of this work was carried out as part of a collaboration with Professor Alessandro Fortunelli and Dr Giovanni Barcaro at the Istituto per i Processi Chimico-Fisici of the CNR (in Pisa, Italy).

Highlights of the Thesis include: the prediction at the DFT level of a novel hybrid structural motif (consisting of a close packed fcc core surrounded by decahedral units) across a wide range of compositions for Pd–Pt clusters of 34 atoms; the prediction that the 98-atom Leary Tetrahedron (LT) structure—while not typically stable for monometallic clusters—may be stabilised for certain bimetallic systems (e.g., Pd–Pt and Au–Pt) again across a wide composition range; the determination of the sensitivity of the structures and chemical ordering of Pd–Pt and Au–Pd nanoalloys on the empirical potential parametrization and how these compare to the results of DFT calculations; the competition between different structural motifs, as a function of elemental composition, for 38-atom clusters of different bimetallic systems; and the investigation of the effect of coordination of CO and H on the relative energies of various structural motifs of 38-atom Pd, Pt and Au clusters.

The work reported in the Thesis will be of primary relevance to those involved in theoretical and experimental studies of nanoalloy cluster structure and heterogeneous catalysis by bimetallic nanoparticles, but it should also be of interest to researchers working on and to other technological applications of nanoalloys, such as in sensors, optics and magnetics.

January 2011   Professor Roy L. Johnston

Acknowledgments

I would like to thank my supervisor Professor Roy Johnston, for his invaluable help, guidance, an infinite patience due to all my travelling (work related) and support throughout my Ph.D. years in Birmingham, UK. Special thanks to the present and past members of the Johnston's Group as well as other Level 2 members: Graham Worth, Gareth, Graham, Ben, Nico, Faye, Andy Logsdail, Ramli, Paul, Cristina, Adam, Dan, Jan, Lasse, Johanna, Tom, Nicola, Heather, Emma, John, and Andy Bennett for all their friendship, and the great times in and out the office. They definitely made my time in Birmingham an unforgettable experience.

Plenty of this work could have not been possible without collaborative work with Professor Alessandro Fortunelli (CNR-Pisa). I would like to thank him for giving me the opportunity to work with him and his group. Thanks to Dr. Giovanni Barcaro (CNR-Pisa) for his invaluable technical help and sincere friendship. To Professor Riccardo Ferrando (Genova), Professor Dominique Bazin (Paris) and Dr. Haydar Arslam (Zonguldak) for helpful discussions. I would also like to acknowledge funding/cpu-time for different sources: CONACYT (Mexico), for the award of a Ph.D. scholarship; the University of Birmingham Overseas Research Student Award (ORSAS-UK); the National Service for Computational Chemistry Software (NSCCS-UK); the HPC-Europa Transnational Access Programme (EU) as well as HPC-Materials Consortium (UK).

On the personal note, I would like to thank my family, my parents (Dora and Alvaro), brother and sister (Alvaro and Atenas) and grandparents (Rosario y Salvador, Feliciano y Maria Loreto), for believing in me and encouranging me to do the best. Special thanks to Marlis, for all her support and patience throughout my Ph.D. From all of them I have learned that, even in difficult times, everything is possible with hard work and patience.

Contents

Chapter 1
Introduction

Metals have always been linked to advances in human civilization. Since the early Bronze Age (approximately 3500 years B.C.), civilization slowly started using metals instead of stones, bones, wood and other non-metal objects as tools. Initially, these primitive tools like knives, axes and arrowheads, were usually made from iron. Metals started being widely used in everyday's life, but also in jewellery, in arts, and ornaments. Precious metals, such as gold and silver, later set the trend for the development of commercial activity as they started being used as coinage [1, 2]. Nevertheless, the discovery of new properties such as malleability, ductility and strength offered by early alloys, e.g. bronze (formed by alloying copper with a small proportion of tin), brass (formed by combining copper and zinc) and later steel (alloying iron with a small amount of carbon), proved to significantly contribute to the improvement of the human quality of life.

The early concept of nanoscience or nanotechnology (structures having sizes of the order of 10^{-9} metres) was first introduced in 1960's by the famous American scientist Richard Feynman, when he asked the question: "how small can we go in the process of miniaturization?", during one of his lectures at Caltech; thus challenging scientists to develop new technologies which could allow the construction and manipulation, at the atomic level, of devices comprising up to a few hundred atoms [3]. Nowadays, Feynman's question is still in the air, and the design of new materials at the nanoscale—not only having novel but well-defined and tunable properties—has attracted an enormous amount of scientific interest. This is further motivated by the wide range of technological applications in which nanoparticles have found their niche: from chemical sensors, optical devices, medicine, surface coatings to cosmetics, among others. Specifically, metal nanoparticles are know to play a crucial role as heterogeneous catalysts, having a key impact on strategic value-adding industries being these foremost the petrochemical, pharmaceutical and clean energy sectors [4, 5].

With modern technological advances on the fabrication,manipulation and visualization techniques at the atomic scale, e.g. electron beam lithography, scanning

L. O. Paz Borbón, *Computational Studies of Transition Metal Nanoalloys*,
Springer Theses, DOI: 10.1007/978-3-642-18012-5_1,
© Springer-Verlag Berlin Heidelberg 2011

tunneling microscopy (STM), scanning transmission electron microscopy (STEM), atomic force microscopy (AFM), high angle annular dark field (HAADF) among others, enourmos progress has been achieved in the field. Furthermore, the development of new and powerful computational resources involving massively parallel and distributed grid computing, along with the implementation of *state-of-the-art* theoretical approaches—mainly based on *first-principles* methodologies—in scientific software, have allowed to carry out sophisticated simulations, unraveling the fundamental physical and chemical properties of nanosystems involving metal particles, hitherto unknown. Such synergy between experiments and computational modelling have clearly proved to be an invaluable tool, not only for shedding light on the interpretation of experimental data, but also for guiding new and novel experiments, efficiently tackling current challenges at the nanoscale [1, 6, 7].

Along these lines, the scientific research documented in this thesis focuses on the computational modelling of transition metal (TM) nanoparticles e.g. Pd–Pt, Ag–Pt, Au–Ag, Pd–Au), involving a few tens of atoms in the gas-phase. In order to achieve this, we used global optimization techniques (such as Genetic Algorithms and Basin Hopping Monte Carlo)—coupled with a semi-empirical Gupta-type potential—for a rapid exploration of the metal nanoalloy potential energy surface (PES), along with Density Functional Theory (DFT) calculations. This "combined approach" allowed us to determine the corresponding global minima (GM) atomic arrangements of these metal nanoparticles and the correct "chemical ordering" (i.e. the way the two metals segregate) at specific cluster sizes and compositions. Moreover, cluster morphologies have been categorized into different *structural families o motifs*: varying from fcc-type crystalline motifs, to decahedral and icosahedral arrangements, to more amorphous configurations. From this analysis, we were thus also able to determine the preference of a particular metal atom to segregate to cluster surface sites (core–shell vs. mixed motifs), for each the nanoalloys described above. Furthermore, three different parametrizations of the Gupta potential for Pd–Au clusters—with direct influence on cluster segregation properties—are studied, as well as the effect of Gupta potential parameters on the structures and segregation in Pd–Pt clusters. Finally, in the last part of this thesis the chemisorption of CO molecules and H atoms on pure 38-atom Pd, Pt and Au, and bimetallic Pd–Pt and Pd–Au clusters, as well as its direct effects on cluster morphology is studied by means of DFT calculations.

1.1 Clusters: Overview and Applications

Clusters can be defined as agglomerates of a few to millions of atoms or molecules. They can be made of one single atom (or molecule) or two or more different species. Due to their size they can have very specific properties and present a wide variety of structures: varying from non-crystalline (e.g. icosahedral), to crystalline arrangements (e.g. fcc-type bulk fragments) as a function of size. They can be studied in the gas-phase ("free-cluster"), passivated byligands (the cluster surface is stabilized by surfactant molecules) or supported on a substrate (such as silica, TiO_2 and MgO) or in an inert gas matrix. One of their main characteristics is that, being so small, they present a high surface/volume ratio. This implies that the nanoparticle has a large number of atoms occupying surface sites [1, 6–8]. These physical properties makes them ideal as catalysts, as it is known that the morphology of the cluster is essential in determining their catalytic properties, with many reactions taking place on nanoparticle surface sites [1, 6–9].

It is interesting to note that, due to their size, metal clusters present properties which make them differ from bulk metals. In the bulk, every metal has filled (or partially filled) electronic bands, with the highest occupied state (at $T = 0$ K) corresponding to the Fermi level (E_F). Above E_F, there is a continuum of energy levels and the electronic occupation of the allowed energy bands give rise to metals insulators (with a large band gap between filled and empty bands); semiconductors (with a band gap comparable to the thermal energy) and metals (with partially filled conduction band), with these properties drastically changing as a function of external factors, e.g. pressure, strain or a magnetic field [10]. Furthermore, these properties can also dramatically change as a function of particle size, i.e. the *size-induced metal-insulator-transition*. This implies that a sucessive fragmentation of a macroscopic piece of a bulk metal will induce a transition to a microscopic insulating particle, with predicting changes in the behaviour of a bulk metal by Kubo for particle's diameter of less than 100Å [11]. It was proposed that, at the microscopic level, the average spacing between discrete electronic energy levels is approximately given by the ratio $\delta = 4E_F/3N$, where N is the total number of valence electrons. In this model, a cluster will have "metallic" properties only when $\delta \ll k_B T$, otherwise the cluster will behave as a semiconductor ($\delta \sim k_B T$) or as an insulator ($\delta \gg k_B T$) [10, 11].

It is difficult to unambiguously define a cluster as being *small*, *medium* or *large* in size. Descriptions often vary, but usually *small clusters* are described as those whose properties are strongly related to their size and morphology. This implies that even a small change in the total number of particles in the cluster will have a significant effect on its fundamental properties, i.e. "every atom counts". For *medium sized clusters* their properties tend to vary more smoothly with size, though there may be some discontinuous behaviour. For *large clusters*, on the other hand, their properties usually resemble those of the bulk material, with a tendency to adopt more pseudo-spherical morphologies [8].

1.2 Nanoalloys

In the last 30 years, a tremendous growth in interest in bimetallic and multimetallic clusters (also know as "nanoalloys") has been seen, due to the fact nanoalloys often present unique characteristics as they are finite-size objects. Novel and interesting nanostructures; which may have fascinating chemical and physical properties could be obtained by having a precise a precise control on the nanoparticle size, composition and degree of "chemical ordering" (degree of mixing) between the metal constituents [1, 6, 7, 12, 13].

Some recent industrial applications of metal nanoparticles and nanoalloys are as heteregenous catalysts. Examples can be found involving global companies like Johnson Matthey Ltd., who have long been using Pt-based honeycomb catalysts for specific Emision-Control-Technologies, i.e. the treatment of exhaust gases (e.g. NO_x and CO) which are generated as combustion products in car engines [14]; as catalysts for chemical reactions in the oil refinery industry as well as novel fuel cells [15]; BASF GmbH (nanotechnology-based products such as self-cleaning coatings, UV-light filtering sunscreens, binders for more weather-resistant exterior paints, etc. [16]); Royal Dutch Shell has current projects in the development of modern Fisher–Tropsch reactions (using iron- and cobalt-based catalysts) to process cleaner fuels (like sulphur-free gasoline and gasoil) from raw natural gas (i.e. Gas-Into-Liquids Technology) [17] among other projects.

1.2.1 Homotops

Compared to monometallic clusters, bimetallic nanoalloys are of greater complexity, due to the presence of two different types of atoms, thus leading to the existence of *homotops*. This term has been introduced by Jellinek [13, 18] to describe alloy cluster isomers $(A_m B_n)$ with fixed number of atoms $(N = m + n)$ and composition $(m/n$ ratio), which have the same structural (geometrical) arrangement but differ in the way in which A and B type atoms are arranged within the particle. The number of homotops in a given cluster is given by the following expression:

$$N_{\text{Homotops}} = \frac{N!}{N_m! N_n!} = \frac{N!}{N_m!(N - N_m)!} \tag{1.1}$$

N_H is maximized when $N_m = N_m$, in other words for 50:50 mixture of elements A and B in the nanoalloy. As an example, for the composition $Pd_{10}Pt_{10}$ there are 184,756 homotops for each geometrical arrangement, while for $Pd_{49}Pt_{49}$ this number increases to approximately 2.5×10^{28} homotops, although some of these may be symmetry-equivalent. In this way, as the cluster size increases, the number of homotops increases combinatorially. This makes the search for global minima (GM) configurations an extremely difficult task, thus highlighting the need for using efficient global optimization schemes.

1.2.2 Segregation in Nanoalloys

The way atoms A and B are arranged in a A-B nanoalloy, i.e. the chemical ordering, will create a characteristic mixing pattern. Figure 1.1 shows different

Fig. 1.1 Different
segregation arrangements
found in nanoalloys (taken
from reference [7])

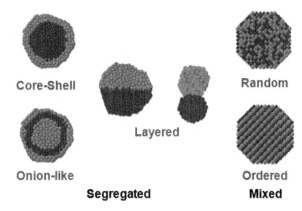

Core-Shell

Layered

Random

Onion-like

Ordered

Segregated

Mixed

possible ways in which atoms A and B can be arranged, according to mixed or segregated behaviour.

Core–shell One element (A) occupies core positions in the nanoparticle, while a shell, created of B atoms, completely surrounds the core. This pattern is denoted as $A_{core}B_{shell}$. This nomenclature is used throughout this thesis.

Onion-like Onion-like configurations are usually found in medium-large size clusters, in which clusters present a layered A-B-A alternating shell pattern (see Chaps. 4 and 7, for onion-like 98-atom Pd–Pt and Pd–Au clusters, respectively).

Layered Layered structures minimize the number of A-B bonds. They are in contact only at a small interface. This pattern ("spherical-cap" segregation) is found in our simulations for 34-atom Pd–Pt clusters for certain potential parameterisations (for more details see Chap. 6).

Random mixing Randomly mixed nanoalloys are usually referred to as "alloyed" nanoparticles, corresponding to bulk solid solutions.

Ordered mixing Pseudo-crystalline regular arrangements of A-B atoms.

There are many factors contributing to these different type of segregation patterns. They depend on properties such as: (a) *bond strength* the relative strengths of A-A, B-B and A-B bonds between metal atoms. If the A-A or B-B bonds are stronger, this favours segregation (the element forming stronger bonds will tend to segregate to core positions), while stronger A-B bonds will favour mixing; (b) *surface energy*, the metal with lower surface energy will tend to segregate to nanoparticle surface sites; (c) *atomic size*, this relates to the release of strain effects in the nanoparticle, with the smaller atom tending to occupy core positions; (d) *charge transfer*, charge transfer between atoms means that the most electronegative atom will tend to occupy surface sites (see Chap. 8, for the specific case of 38-atom Ag–Au); (e) *strength of binding to ligands*, for passivated clusters, the metal which binds more strongly to the ligand, will prefer to occupy surface sites (see Chap. 8 for the specific case of Pd–Au and Pd–Pt clusters interacting with 32 CO molecules); (f) *specific electronic effects*, for certain metals, may play an important role in determining segregation (directionality of bonding).

From a theoretical point of view, a challenging problem in cluster science is to determine the most stable cluster (GM), for a given size and composition, in terms of geometry and chemical ordering as well as electronic structure. [19–26]. Once this problem is solved, structure-property relationships can be investigated.

1.3 Experimental Background: From Synthesis to Cluster Characterisation

So far we have described the many fundamental properties, characteristics and technological applications of nanoparticles. It is also important to outline the experimental synthesis (production) and characterisation (visualization) of nanoparticles, though this research has focused only on the theoretical/computational modelling of nanoalloys. One of the most commonly used experimental techniques for the generation of gas-phase clusters is the cluster molecular beam (see Fig. 1.2).

First, a small metal target (plate or rod) of *bulk material* is *evaporated* by laser ablation (Fielicke 2008, personal communication). For the case of bimetallics nanoalloys, a single alloy target may be used, or two pure metal targets. Evaporation can also be achieved by other methods such as *heavy ion sputtering*, *magneton sputtering* or *electrical discharge*. This generates a plasma which is cooled via collisions with a cold carrier inert gas (usually He or Ar), which results in condensation and *cluster nucleation*. Clusters *grow* by collisions. Adiabatic and isenthalpic expansion leads to strong *cooling*, as the clusters undergo a *supersonic expansion*, as they pass from the high-pressure *condensation* region through a narrow nozzle into a vacuum. Within the *supersonic jet*, there are no collisions and the properties of isolated clusters can therefore be studied (see Fig. 1.2) [7, 27 and Fielicke (2008, personal communication)].

Other chemical techiques for cluster generation, such as *chemical reduction*, in which clusters are produced by reduction of solutions metal salts, dissolved in solvent, in the presence of surfactant. Particle size can be controlled by the use of inverse micelles; and *electrochemical synthesis*, where metals such as Pt, Rh and Ru can be generated by reducing their corresponding salts at the cathode [7].

For cluster characterization, several experimental techniques are used, varying from *mass spectroscopy*, in which an ionized cluster beam is deflected by applying an electric field. By tuning the intensity of the electric field, one can "select"

Fig. 1.2 Experiments under collision-free conditions [27]

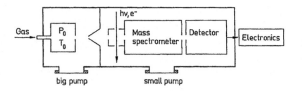

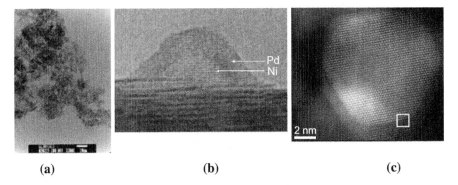

Fig. 1.3 Electron microscopy images of supported nanoalloys. **a** TEM image of Pd_6Pt catalyst (~ 6 nm); **b** HRTEM image of Ni-Pt nanoparticle (~ 3 nm) and **c** HAADF-STEM image of a three-layer cuboctahedral Pd–Au nanoparticle (~ 12 nm) [28], [29] (Figure (c) [30] Reproduced by permission of the *Royal Society of Chemistry*)

cluster mass, i.e. cluster particle size; to *X-ray diffraction*, for determining the structure and crystallinity of the cluster. *Electron Microscopy techniques* are also widely used, such as: scanning transmission electron microscopy (STEM); High-Resolution TEM (HRTEM), Z-contrast High Angle Annular Dark Field (HAADF-STEM) is used to reveal the internal structure of the nanoparticle. *Scanning Probe Microscopy*, such as Atomic Force Microscopy (AFM) and Scanning-Tunneling Microscopy (STM); with the rastering of a sharp tip across the nanoparticle surface. The strength of this interaction is use to picture the topography of the nanoparticle. *X-Ray spectroscopy techniques*, such as X-Ray Absorption Spectroscopy (XAS) and Extended X-Ray Absorption Fine Structure (EXAFS) are based on the principle that each element has a unique X-ray absorption spectrum; so each element in the nanoparticle can be identified (electron microscopy images of supported nanoalloys are shown in Fig. 1.3).

1.4 Second and Third Rows Transition Metal Nanoalloys

We have chosen to study the four second and third rows nanoalloy systems, i.e. Pd–Pt, Pd–Au, Ag–Pt and Ag–Au. These systems are described briefly below and further details can be found in the relevant results chapters.

1.4.1 Pd–Pt Clusters

In the bulk, Pd and Pt form solid solutions (i.e. they mix) for all compositions [31]. Alloying Pd and Pt atoms is claimed to be more catalytically active for aromatic hydrocarbon hydrogenation and more resistant to sulfur poisoning than either of the pure metals, due to "synergistic" effects [32], though there is some controversy over this viewpoint [33, 34]. Renouprez and co-workers have performed extensive experimental studies of the structures, compositions, and catalytic activities of Pd–Pt particles generated by the laser vaporization of bulk alloys of various compositions [33, 34]. Their results, obtained from a combination of experimental techniques (TEM, EXAFS and Low Energy Ion Scattering), indicate that the Pd–Pt nanoparticles (1–5 nm in diameter) are truncated octahedra with a Pt-rich core surrounded by a Pd-rich shell [33, 34]. In more recent studies by Bazin and co-workers, Pd–Pt nanoalloys were deposited on γ-alumina and by combining X-ray absorption spectroscopy with transmission electron microscopy (TEM) and volumetric H_2-O_2 titration, it was shown that small Pd–Pt particles (~ 1 nm) have "cherry-like" structures with a distribution of Pd atoms on the surface of the cluster [28].

Theoretical studies of Pd–Pt nanoparticles have been performed by Massen et al., using a combination of a genetic algorithm and the Gupta potential [20]. Their results showed that Pd–Pt particles adopt different morphologies compared to the pure Pt and Pd clusters. Pd surface segregation was predicted and was explained in terms of the lower surface energy of Pd, and the higher cohesive energy by Pt. These results are in agreement with DFT calculations performed by Fernandez et al., for which $Pt_{core}Pd_{shell}$ segregation was found to be the most stable configuration for $(Pd–Pt)_N$ clusters, with $N = 5 - 22$ [35].

1.4.2 Pd–Au Clusters

The catalytic properties of Pd–Au nanoparticles are due to their eletronic structures which differ from pure Au and Pd nanoclusters. It has been found that Pd–Au nanoalloys are efficient catalysts for a wide variety of chemical reactions, such as: acetylene cyclotrimerization; the hydrogenation of hex-2-yne to cis-hex-2-ene; hydrodechlorination of trichloroethene (TCE) in water; low temperature synthesis of hydrogen peroxide; and the reduction of CO and alcohols [36–39].

Previous experimental studies of Pd–Au nanoalloys have shown that both $Pd_{core}Au_{shell}$ and $Au_{core}Pd_{shell}$ configurations can be generated by different preparation methods [37, 40, 41], while combined experimental/theoretical studies have indicated the preference for $Pd_{core}Au_{shell}$ (~ 5 nm) and an "onion-like" Pd–Au–Pd configuration for larger Pd–Au nanoalloys (~ 12 nm) [30, 42, 43].

1.4.3 Ag–Pt Clusters

Experimental studies of Ag–Pt nanoparticles, prepared by radiolysis, suggested nearly spherical morphologies [44]. Ag–Pt nanowires (lengths $\sim 3.5\mu$, diameters 3–20 nm) have also been studied by Doudna et al. They reported polycrystalline

Pt–Ag nanowires with the grains having fcc-type packing [45]. Their EXAFS analysis (Pd L3 edge) showed little Pt–Ag mixing, with Ag occupying core positions and Pt forming a surface shell, resulting in $Pt_{shell}Ag_{core}$ structure.

1.4.4 Ag–Au Clusters

Ag and Au are known to form solid solutions in the bulk (all compositions). It it also know that the mixing (in the bulk) of Ag and Au is weakly exothermic, with little surface segregation for Ag–Au alloys. Ag–Au nanoparticles have been widely studied because of their optical properties. Of particular interest is the shape and frequency of the plasmon resonance as a function of composition as well as segregation or mixing of the Ag–Au nanoparticle [46]. Related to catalytic applications, Ag–Au nanoparticles are not as widely used as Pt-based systems, although they are used specifically for alkene epoxidation [47].

Ag–Au nanoparticles have been generated by chemical or electrochemical deposition of one metal onto a preformed cluster of the other, yielding both $Ag_{core}Au_{shell}$ and $Au_{core}Ag_{shell}$ configurations [48, 49]. They have also been prepared by the method of reduction from a solution containing a mixture of Ag–Au salts, by Han et al. [50]. In their study, FT-IR spectroscopy revealed that the nanoparticle surfaces were enriched in Ag.

1.5 Chemisorption on Single Metal and Bimetallic Clusters

An important difference of studying (both experimentally and computationally) chemical reactions on a finite cluster, rather than on a periodic infinite surface, is that clusters often present chemically active coordination sites which are not present on an ideal extended surface (Fig. 1.4 illustrates different models in theoretical cluster science for studying chemisorption processes). The nature and number of active surface sites will vary with particle shape, packing and size. For the case of bimetallic clusters, there will also be a strong dependence on the elemental composition and the degree of segregation or mixing of the component metals.

In recent experiments using XASX-ray absorption spectroscopy (XAS), Bazin have demonstrated the link between the adsorption mode of NO molecules and the behaviour of the metal cluster after adsorption [55]. In the case of dissociative adsorption, they found that metal oxide clusters are formed (decreasing the catalytic activity), whereas non-dissociative molecular adsorption (in the high temperature regime) leads to sintering of the metallic cluster. Bazin and co-workers have proposed that the nature of the adsorption mode of NO molecules on metallic clusters can either be stable under this adsorption, where sintering of the metallic cluster can occur, or it may undergo a disruption of the metal-metal bonds (see Fig. 1.5).

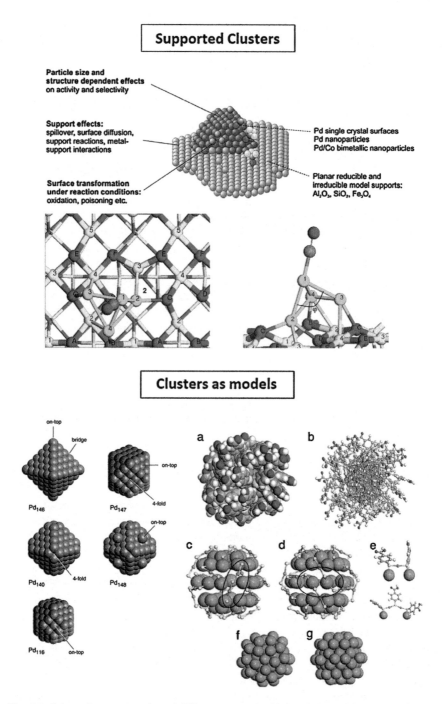

Fig. 1.4 Schematic representation of different approaches for studying nanoparticles and their interactions with small molecules, [51, 52]. Ref. [53], "Copyright (2008) National Academy of Sciences, U.S.A". (Figure from Ref. [54]—Reproduced by permission of the *Royal Society of Chemistry*)

Fig. 1.5 Correlation diagram between adsorption mode and the behaviour of metallic clusters, i.e. dissociative adsorption $(N_a + O_a)$ with cluster fragmentation occurs above the straight line, while associative adsorption (NO_a) with cluster sintering occurs below this line, as proposed by Bazin [55]

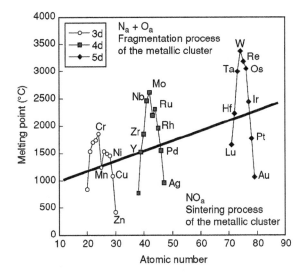

Burke et al. performed studies based on temperature programmed desorption (TPD) and analysed the influence of sulfur on the adsorption of H_2 on the Pd(100) surface. They found that, after reaching a specific sulfur concentration on the Pd surface (above 0.28 ML), no hydrogen adsorption was detected [56]. Following studies of hydrogen adsorption kinetics on Pd(100) surfaces in the presence of sulfur impurities, Peden et al. reported that the adsorption rate of hydrogen is strongly attenuated by the presence of sulfur impurities, suggesting that sulfur-poisoned surfaces drastically alter the H_2 adsorption process [57]. DFT calculations by Gravil et al. on hydrogen adsorption on Pd(111) surfaces, showed that adsorption energies decrease while diffusion barriers increase for hydrogen atoms in the vicinity of sulfur adatoms. After exceeding a concentration of 0.33 ML sulfur atom coverage, full poisoning of the hydrogen adsorption on Pd(111) was observed [58].

In is manner, the combination of novel computational and theoretical approaches are crucial in order to underpin experimental work. *State-of-the-art* ab initio methodologies, such as DFT, have shown to play a key role in understanding the electronic structure of metal nanoparticles and determining their corresponding physical and chemical properties. These are clearly the very first steps towards a deeper understanding of catalysis by them.

References

1. F. Baletto, R. Ferrando, Rev. Mod. Phys. **77**, 371 (2005)
2. J. Jellinek, Faraday Discuss. **138**, 11 (2008)
3. R. Feynam, There's plenty of room at the bottom, was originally published in the February 1960 issue of Caltech's Engineering and Science Magazine (1960)

4. H.-J. Freund, Surf. Sci. **500**, 271 (2002)
5. J.K. Norskov,T. Bligaard, J. Rossmaisl, C. Christensen, Nat. Chem. **1**, 37 (2009)
6. R.L. Johnston, *Atomic and Molecular Clusters*, (Taylor and Francis, London 2002)
7. R. Ferrando, R.L. Johnston, J. Jellinek, Chem. Rev. **108**, 845 (2008)
8. H. Haberland (ed.), *Clusters of Atoms and Molecules*, vols. I and II. (Springer, Berlin, 1994)
9. R.L. Johnston, R. Ferrando (eds.) Nanoalloys: from theory to applications. Faraday Discuss **138**, 1–441 (2008)
10. C. Kittel, *Introduction to Solid State Physics Eight Edition*, (Wiley, USA, 2005)
11. P. Braunstein, L.A. Oro, P.R. Raithby (eds.), *Metal Clusters in Chemistry*, vols. 1–3. (Wiley-VCH, Weinheim, 1999)
12. P.K. Jain, I. El-Shayed, M.A. El-Shayed, Nanotoday **2**, 18 (2007)
13. J. Jellinek (ed.), *Theory of Atomic and Molecular Clusters*. (Springer, Berlin, 1999)
14. See Johnson Matthey Technology Centre http://www.matthey.com/about/envtech.htm, Sonning Common, United Kingdom
15. See Haldor Topsoe Research and Development, http://www.topsoe.com/Researchingcatalysis. aspx
16. See GmbH. BASF, Nanotechnology in Dialogue, http://www.basf.com/group/corporate/ en/function/conversions:/publish/content/innovations/events-presentations/nanotechnology/ images/dialog.pdf
17. See Royal Dutch Shell, http://realenergy.shell.com/
18. E.B. Krissinel, J. Jellinek, Int. J. Quant. Chem. **62**, 185 (1997)
19. S. Darby, T.V. Mortimer-Jones, R.L. Johnston, C. Roberts, J. Chem. Phys. **116**, 1536 (2002)
20. C. Massen, T.V. Mortimer-Jones, R.L. Johnston, J. Chem. Soc. Dalton Trans. 4375 (2002)
21. L.D. Lloyd, R.L. Johnston, S. Salhi, N.T. Wilson, J. Mater. Chem. **14**, 1691 (2004)
22. L.D. Lloyd, R.L. Johnston, S. Salhi, J. Comp. Chem. **26**, 1069 (2005)
23. L.O. Paz-Borbón, R.L. Johnston, G. Barcaro, A. Fortunelli, J. Phys. Chem. C **111**, 2936 (2007)
24. L.O. Paz-Borbón, R.L. Johnston, G. Barcaro, A. Fortunelli, J. Chem. Phys. **128**, 134517 (2008)
25. L.O. Paz-Borbón, A. Gupta, R.L. Johnston, J. Mater. Chem. **18**, 4154 (2008)
26. F. Pittaway, L.O. Paz-Borbón, R.L. Johnston, H. Arslan, R. Ferrando, C. Mottet, G. Barcaro, A. Fortunelli, (submitted, 2009)
27. A. Fielicke, Reactivity and Catalysis. http://www.fhi-berlin.mpg.de/mp/fielicke/ (2008)
28. D. Bazin, D. Guillaume, C.H. Pichon, D. Uzio, S. Lopez, Oil Gas. Sci. Tech. Rev. IFP **60**, 801 (2005)
29. S. Sao-Joao, S. Giorgio, J.M. Penisson, C. Chapon, S. Bourgeois, C. Henry, J. Phys. Chem. B **109**, 342 (2005)
30. D. Ferrer, D. Blom, L.F. Allard, S.J. Mejia-Rosales, E. Pérez-Tijerina, M. José-Yacamán, J. Mater. Chem. **18**, 2442 (2008)
31. F.R. De Boer, R. Boom, W.C.M. Mattens, A.R. Miedama, A.K. Niessen, *Cohesion in Metals: Transition Metal Alloys* (Elsevier, Amsterdam, 1988)
32. A. Stanislaus, B.H. Cooper, Catal. Rev. Sci. Eng. **36**, 75 (1994)
33. A.J. Renouprez, J.L. Rousset, A.M. Cadrot, Y. Soldo, L. Stievano, J. Alloy. Comp. **328**, 50 (2001)
34. J.L. Rousset, L. Stievano, F.J. Cadete Santos Aires, C. Geantet, A.J. Renouprez, M. Pellarin, J. Catal. **202**, 163 (2001)
35. E.M. Fernández, L.C. Balbás, L.A. Pérez, K. Michaelian, I.L. Garzón, Int. J. Mod. Phys. **19**, 2339 (2005)
36. A.F. Lee, C.J. Baddeley, C. Hardacre, R.M. Ormerod, R.M. Lambert, G. Schmid, H. West, J. Phys. Chem. **99**, 6096 (1995)
37. G. Schmid, in *Metal Clusters in Chemistry*, vol. 3, eds. by P. Braunstein, L.A. Oro, P.R. Raithby (Wiley-VCH, Weinheim, 1999) p. 1325
38. M.O. Nutt, J.B. Hughes, M.S. Wong, Environ. Sci. Technol. **39**, 1346 (2005)

39. J.K. Edwards, B.E. Solsona, P. Landon, A.F. Carley, A. Herzing, C.J. Kiely, G.J. Hutchings, J. Catal. **236**, 69 (2005)

40. H. Remita, M. Mostafavi, M.O. Delcourt, Radiat. Phys. Chem. **47**, 275 (1996)

41. K. Luo, T. Wei, C.-W. Yi, S. Axnanda, D.W. Goodman, J. Phys. Chem. B **109**, 23517 (2005)

42. S.J. Mejia-Rosales, C. Fernandez-Navarro, E. Pérez-Tijerina, D.A. Blom, L.F. Allard, M. José-Yacamán, J. Phys. Chem. C **111**, 1256 (2007)

43. M. José-Yacamán, S.J. Mejia-Rosales, E. Pérez-Tijerina, J. Mater. Chem. **17**, 1035 (2007)

44. M. Treguer, C. de Cointet, S. Remita, M. Khatouri, M. Mostafavi, J. Amblard, J. Belloni, J. Phys. Chem. B **102**, 4310 (1998)

45. C.M. Doudna, M.F. Bertino, F.D. Blum, A.T. Tokuhiro, D. Lahiri-Dey, S. Chattopadhay, J. Terry, J. Phys. Chem. B **107**, 2966 (2003)

46. U. Kreibig, M. Quinten, in *Clusters of Atoms and Molecules*, vol. II, ed. by H. Haberland (Springer, Berlin, 1994), p. 321

47. N. Toreis, X.E. Verykios, Surf. Sci. **197**, 415 (1988)

48. J.H. Hodak, A. Henglein, M. Giersig, G.V. Hartland, J. Phys. Chem. B **104**, 11708 (2000)

49. J.P. Wilcoxon, P.P. Provencio, J. Am. Chem. Soc. **126**, 6402 (2004)

50. S.-W. Han, Y. Kim, K. Kim, J. Colloid Interface Sci. **208**, 272 (1998)

51. M. Valero Corral, P. Raybaud, P. Sautet, J. Catal. **247**, 339 (2007)

52. I.V. Yudanov, R. Sahnoun, K.M. Neyman, N. Rösch, J. Hoffmann, S. Schauermann, V. Johnek, H. Unterhalt, G. Rupprechter, J. Libuda, H.-J. Freund, J. Phys. Chem. B **107**, 255 (2003)

53. M. Walter, J. Akola, O. Lopez-Acevedo, G. Calero, C.J. Ackerson, R.L. Whetten, H. Grönbeck, H. Häkkinen, PNAS **105**, 9157 (2008)

54. M. Bäumer, J. Libuda, K.M. Neyman, N. Rösch, G. Rupprechter, H.-J. Freund, Phys. Chem. Chem. Phys. **9**, 3541 (2007)

55. D. Bazin, Macro. Res. **17**, 230 (2006)

56. M.L. Burke, R.J. Madix, Surf. Sci. **237**, 1 (1990)

57. C.H.F. Peden, B.D. Kay, D.W. Goodman, Surf. Sci. **175**, 215 (1986)

58. P.A. Gravil, H. Toulhoat, Surf. Sci. **430**, 176 (1999)

Chapter 2
Theoretical Background and Methodology

In this chapter the different approaches followed in order to model metal nanoparticles are described. The theoretical framework adopted in this work, varying from the use of an empirical potential (i.e. the Gupta potential) to ab initio methods such as density functional theory (DFT) is explained. An introduction to different global optimization techniques, for the exploration of the nanoparticle potential energy surface, as well as a combined Gupta-DFT approach is also given. Finally, energetic quantities for analysing the stability of nanoparticles are described in detail.

2.1 Modelling Metal Clusters and Nanoalloys

Different approaches can be used to model atom–atom interactions in metal clusters and nanoalloys, and to search for putative stable configurations, i.e. to locate the global minimum (GM) on the nanoparticle potential energy surface (PES). This is a very difficult task because the properties of each system will depend on its constituent elements and size. The complexity will also depend on how accurate the model we are using is, ranging from: pair-wise potentials (such as the Lennard–Jones potential; an approximation which is quite accurate for noble gases, in which the interatomic interactions depend mainly on the distances between atoms); and many-body potentials (e.g. the Gupta potential, Embedded Atom and Sutton–Chen potentials); to computationally intensive *first-principle* methodologies (e.g. Hartree–Fock calculations, DFT, or more expensive post-Hartree–Fock methods such as configuration interaction (CI) or Møller–Plesset perturbation theory). Pair-wise potentials are simple approximations for describing atomic interactions, in which the nature of the interactions depends on attractive and repulsive terms (having parameters, fitted either to experimental observations or calculated theoretically); and where the total

L. O. Paz Borbón, *Computational Studies of Transition Metal Nanoalloys*,
Springer Theses, DOI: 10.1007/978-3-642-18012-5_2,
© Springer-Verlag Berlin Heidelberg 2011

energy of the system can be calculated in terms of the atomic positions $(r_1, r_2, \ldots, r_N)$. Many-body potentials, such as Gupta, tend to overcome the high computational cost/effort of first principles calculations while keeping the many-body nature of metallic bonding. When coupled with global optimization tools (e.g. Genetic algorithms and Basin Hopping Monte Carlo algorithms) this allows us to explore large areas of the nanoparticle PES in a feasible amount of time, while at the same time, accurately simulating the interatomic interactions.

2.1.1 Potential Energy Surfaces

One of the main goals in this study is to determine the global minimum (GM) structure (i.e. the structural configuration with the lowest total potential energy) for a metal cluster of a certain size and elemental composition. It is known that experimental information is not always sufficient for determining the nanocluster structure precisely and theoretical predictions are of great importance in this field. Understanding the structural configuration of nanoclusters (as well as their energetics) will aid the tailored design of nanoparticles, in which the nanoparticle's physical and chemical properties can be fine tuned. The potential energy of a nanoparticle (V_{clus}), can be represented on a potential energy surface (PES) diagram. The PES of a system is represented in terms of the atomic coordinates. The number of interacting atoms in the system (N) leads to $3N$ degrees of freedom, yielding a PES dimensionality of $3N + 1$, where the extra dimension is the potential energy.

A local minimum in the PES is defined as a point, in which any displacement will lead to higher potential energy (V_{clus}) configurations. The gradients at this point are all zero, $\nabla V_{\text{clus}} = 0$, and all the curvatures (second derivatives) are positive. The PES of a small system can have large numbers of *local-minima*, which correspond to high-energy arrangements. The lowest energy configuration is called the *global-minumum* (GM) [1]. Figure 2.1 shows a schematic representation of two systems of fixed size and composition, but having differ segregation patterns between A (blue) and B (grey) metals. One can see that a particular chemical ordering in the nanoparticle will lead to a lower energy configuration (a more stable structure).

2.2 Theoretical Methods

2.2.1 The Gupta Potential

The Gupta potential [2, 3] is based on the second moment approximation to Tight Binding theory. It is written in terms of repulsive (V^r) pair and attractive many-body (V^m) terms, which are obtained by summing over all (N) atoms:

Fig. 2.1 Pictorical representation of a potential energy surface of two bimetallic cluster homotops, having the same number of atoms A (*grey*) and B (*blue*) and geometries. They occupy different regions in the PES (different *basins*), due to differences in *chemical ordering*

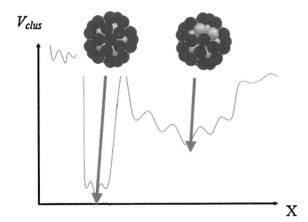

$$V_{\text{clus}} = \sum_i^N \{V^{\text{r}}(i) - V^{\text{m}}(i)\} \tag{2.1}$$

where:

$$V^{\text{r}}(i) = A(a,b) \sum_{j \neq i}^N \exp\left(-p(a,b)\left(\frac{r_{ij}}{r_0(a,b)} - 1\right)\right) \tag{2.2}$$

and

$$V^{\text{m}}(i) = \zeta(a,b)\sqrt{\sum_{j \neq i}^N \exp\left(-2q(a,b)\left(\frac{r_{ij}}{r_0(a,b)} - 1\right)\right)} \tag{2.3}$$

In Eqs. 2.2 and 2.3, r_{ij} represents the interatomic distance between atoms i and j. A, r_0, ζ, p and q are fitted to experimental values of the cohesive energy, lattice parameters and independent elastic constants for crystal structures of pure metals and bulk alloys and a and b define the element types of atoms i and j, respectively. Gupta potential parameters used in this research are listed in Table 2.1. For the discussion of varying Pd–Pt parameters and two different fitted Pd–Au parameter sets, see Chaps. 6 and 7, respectively.

2.2.2 Early Density Functional Theory: Thomas Fermi Model

A first approximation to studies at the atomic level was made using Newtonian mechanics and classical electromagnetism. Due to the failure to explain the structure and complexity of atomic systems, new theories had to be proposed. Thomas and Fermi gave the first formal derivation of a density functional approach for a system

Table 2.1 Gupta potential parameters used in this research [3]

Parameters	Pd–Pd	Pt–Pt	Ag–Ag	Au–Au	Au–Au (for Ag–Au)	Ag–Au	Ag–Pt	Pd–Au	Pd–Pt
A (eV)	0.1746	0.2975	0.1031	0.2061	0.2096	0.1488	0.175	0.19	0.23
ζ (eV)	1.718	2.695	1.1895	1.790	1.8153	1.4874	1.79	1.75	2.2
p	10.867	10.612	10.85	10.229	10.139	10.494	10.73	10.54	10.74
q	3.742	4.004	3.18	4.036	4.033	3.607	3.57	3.89	3.87
r_0(Å)	2.7485	2.7747	2.8921	2.884	2.885	2.8885	2.833	2.816	2.76

of electrons in an external potential, due to nuclei [4, 5]. In their model the total energy of an inhomogeneous electron gas is written as a functional of the electronic density, $\rho(\vec{r})$. This was an initial (rough) approximation to the exact solution of the many-electron Schrödinger equation and the associated wave function $\Psi(r_1, r_2, \ldots, r_N)$, and is the most primitive version of DFT.

$$E_{TF}[\rho(\vec{r})] = T_{TF}[\rho(\vec{r})] + E_{en}[\rho(\vec{r})] + E_{ee}[\rho(\vec{r})] \quad (2.4)$$

In the Thomas–Fermi (TF) model, the electronic density $\rho(\vec{r})$ completely characterizes the ground state of the system. As the TF model neglects exchange (included in the Hartree–Fock method, as a consequence of the anti-symmetry of the wave function, i.e. the Pauli exclusion principle) and correlation effects; the energy functional $E_{TF}[\rho(\vec{r})]$ is expressed as the contribution of kinetic energy $(T_{TF}[\rho(\vec{r})])$, the electron–nucleus attraction $(E_{en}[\rho(\vec{r})])$ and electron–electron repulsion $(E_{ee}[\rho(\vec{r})])$. In the TF model, the kinetic energy is expressed as a functional of the electronic density $\rho(\vec{r})$.

$$T_{TF}[\rho(\vec{r})] = C_F \int \rho^{5/3}(\vec{r}) d\vec{r} \quad (2.5)$$

hence, the energy functional can be written as follows:

$$E_{TF}[\rho(\vec{r})] = C_F \int \rho^{5/3}(\vec{r}) d\vec{r} - Z \int \frac{\rho(\vec{r})}{r} d\vec{r} + \frac{1}{2} \int \int \frac{\rho(\vec{r})\rho(\vec{r}')}{|\vec{r} - \vec{r}'|} d\vec{r} d\vec{r}' \quad (2.6)$$

with $C_F = \frac{3}{10}(3\pi^2)^{2/3} = 2.871$; calculated from the jellium model. The TF model assumes that the electronic density $\rho(\vec{r})$ minimizes the energy functional $E_{TF}[\rho(\vec{r})]$, with the constraint:

$$N = N[\rho(\vec{r})] = \int \rho(\vec{r}) d\vec{r} \quad (2.7)$$

where integrating over $\rho(\vec{r})$, gives the total number of electrons, N. Using the Lagrange multipliers method in order to find a stationary point for $E[\rho(\vec{r})]$, using the constraint 2.7:

$$\frac{\delta}{\delta\rho(\vec{r})}\left[E_{TF}[\rho(\vec{r})] - \mu_{TF}\left(\int \rho(\vec{r}) d\vec{r} - N\right)\right] = 0 \quad (2.8)$$

the variations in electronic density of the energy functional $E_{TF}[\rho(\vec{r})]$, can be expressed as an Euler–Lagrange equation, where:

$$\frac{\delta E_{TF}[\rho(\vec{r})]}{\delta \rho(\vec{r})} = \mu_{TF} \tag{2.9}$$

$$\mu_{TF} = \frac{\delta E_{TF}[\rho(\vec{r})]}{\delta \rho(\vec{r})} = \frac{5}{3} C_F \rho^{2/3}(\vec{r}) - \phi(\vec{r}) \tag{2.10}$$

where $\phi(\vec{r})$:

$$\phi(\vec{r}) = \frac{Z}{r} - \int \frac{\rho(\vec{r'})}{|\vec{r} - \vec{r'}|} d\vec{r} \tag{2.11}$$

Equation 2.11 can be solved using the density constraint (Eq. 2.7), and the resulting density is then inserted in Eq. 2.6, in order to give the total energy of the system. The TF model is a very simple theory for describing total energies of atoms, which set up the foundations for a more complex DFT [6, 7].

2.2.3 Modern Density Functional Theory

For the case of N interacting electrons (in the ground-state), their interactions can be described as the sum of the kinetic energy ($\hat{T}$), external potential ($\hat{V}$) and Coulombic ($\hat{V}_{ee}$) operators. The corresponding Hamiltonian operator ($\hat{H}$):

$$\hat{H} = \hat{T} + \hat{V} + \hat{V}_{ee} \tag{2.12}$$

which can be expressed, in a similar way, in the following analytic form:

$$\hat{H} = -\sum_{i=1}^{N} \frac{1}{2} \nabla_i^2 + \sum_{i=1}^{N} v(\vec{r}_i) + \frac{1}{2} \sum_{i=1}^{N} \sum_{i \neq j}^{N} \frac{1}{|\vec{r}_i - \vec{r}_j|}$$

where $v(\vec{r}_i)$ represents the external potential. The Hamiltonian $\hat{H}$, described in the equation above, takes into account the adiabatic Born–Oppenheimer approximation. The Born–Oppenheimer approximation treats the atomic nucleus as being fixed in position, with respect to the electronic motion, due to large mass differences between the atomic nucleus and the electron. This represents a decoupling of the motion of the electrons and the motion of the nucleus; hence one needs to solve the Schrödinger equation only for the electronic part [8].

The foundations of modern DFT were published in the classic papers of Hohenberg and Kohn in 1964, and Kohn and Sham one year later [9, 10]. They developed an exact variational principle formalism in which ground state properties, such as: total electronic energy, equilibrium positions and magnetic moments are expressed in terms of the total electronic density $\rho(\vec{r})$. The theorems are explained below:

Hohenberg and Kohn Theorem: the ground state density $\rho(\vec{r})$ of a bound system of interacting electrons in some external potential $v(\vec{r})$ determines this potential uniquely, as well as the ground state wave-function $\Psi(r_1, \ldots, r_N)$.

This theorem also states that: since electronic density $\rho(\vec{r})$ determines both the total number of electrons, N and the external potential $v(\vec{r})$; $\rho(\vec{r})$ also provides a full description of all the ground-state observables, which are functionals of $\rho(\vec{r})$ [11]. Another statement of the Hohenberg–Kohn theorem states: the ground state energy E_0 and the ground-state density $\rho_0(\vec{r})$ of a system characterized by an external potential $v(\vec{r})$ can be obtained by using the variational principle (involving only the density).

In other words, the ground state energy E_0 can be expressed as a functional of the ground state density $E_v[\rho(\vec{r})]$, otherwise the inequality prevails [6]:

$$E_0 = E_v[\rho_0(\vec{r})] < E_v[\rho(\vec{r})] \tag{2.14}$$

The theory also established the existence of a universal functional, $F[\rho(\vec{r})]$ which is independent of the external potential $v(\vec{r})$, i.e. which has the same functional form for any system considered. The universal functional $F[\rho(\vec{r})]$, can be expressed as the sum of the kinetic energy of the gas of non-interacting Kohn–Sham electrons ($T_s[\rho(\vec{r})]$), the Coulombic electron–electron interaction, and the exchange-correlation term ($E_{xc}[\rho(\vec{r})]$) as follows:

$$F[\rho(\vec{r})] = T_s[\rho(\vec{r})] + \frac{1}{2} \int \int \frac{\rho(\vec{r})\rho(\vec{r}')}{|\vec{r} - \vec{r}'|} + E_{xc}[\rho(\vec{r})] \tag{2.15}$$

The $E_{xc}[\rho(\vec{r})]$ term is expressed as the sum of the correlation ($E_c[\rho(\vec{r})]$) and the exchange energy ($E_x[\rho(\vec{r})]$). In the $E_{xc}[\rho(\vec{r})]$ term, the correlation energy $E_c[\rho(\vec{r})]$ is expressed as the difference between the true kinetic energy $T[\rho(\vec{r})]$ and $T_s[\rho(\vec{r})]$:

$$E_c[\rho(\vec{r})] = T[\rho(\vec{r})] - T_s[\rho(\vec{r})] \tag{2.16}$$

The exchange $E_x[\rho(\vec{r})]$ term is derived from Hartree–Fock Theory (Slater determinant). The Slater determinant, which takes into account the antisymmetry of the wavefunction from the Pauli exclusion principle (i.e. an electron having either spin α or spin β) assures a change of sign under electron exchange:

$$\psi(\vec{r_1}, \vec{r_2}, \ldots, \vec{r_N})$$
$$= \sqrt{\frac{1}{N!}} \begin{vmatrix} \phi_0(\vec{r_1})\alpha(s_1) & \phi_0(\vec{r_2})\beta(s_1) & .. & \phi_{N/2-1}(\vec{r_1})\alpha(s_1) & \phi_{N/2-1}(\vec{r_1})\beta(s_1) \\ .. & .. & .. & .. & .. \\ \phi_0(\vec{r_N})\alpha(s_N) & \phi_0(\vec{r_N})\beta(s_N) & .. & \phi_{N/2-1}(\vec{r_N})\alpha(s_N) & \phi_{N/2-1}(\vec{r_N})\beta(s_N) \end{vmatrix}$$
$$\tag{2.17}$$

Therefore, the total energy of a system ($E_v[\rho(\vec{r})]$), under a external potential, $v(\vec{r})$, can be expressed as:

$$E_v[\rho(\vec{r})] = F[\rho(\vec{r})] + \int \rho(\vec{r})v(\vec{r})d^3\vec{r} \tag{2.18}$$

Using the variational principle with the constraint 2.7 the ground state density satisfies the stationary principle:

$$\delta\left[E_v[\rho(\vec{r})] - \mu\left(\int \rho(\vec{r})d^3\vec{r} - N\right)\right] = 0 \tag{2.19}$$

which gives the Euler–Lagrange equation:

$$\frac{\delta E_v[\rho(\vec{r})]}{\delta\rho(\vec{r})} = v(\vec{r}) + \frac{\delta F[\rho(\vec{r})]}{\delta\rho(\vec{r})} \tag{2.20}$$

From the Hohenberg–Kohn theorems, we have derived the Kohn–Sham (KS) equations [6, 7, 10]. The KS equations solve the problem of the complex many-electron Schrödinger equations, by transforming them into a set of N single-electron equations, which need to be solved self-consistently.

$$\left[-\frac{1}{2}\nabla^2 + v(\vec{r}) + \int \frac{\rho(\vec{r})}{|\vec{r} - \vec{r}'|}d^3\vec{r}' + v_{xc}[n(\vec{r})]\right]\psi_i(\vec{r}) = \epsilon_i\psi_i(\vec{r}) \tag{2.21}$$

with an electronic density:

$$\rho(\vec{r}) = \sum_{i=1}^{N} |\psi_i(\vec{r})|^2 \tag{2.22}$$

In Eq. 2.21, the exchange-correlation potential, $v_{xc}[n(\vec{r})]$ is expressed as the partial derivative of the $E_{xc}[\rho(\vec{r})]$ term, with respect to the electronic density, $\rho(\vec{r})$:

$$\frac{\delta E_{xc}[\rho(\vec{r})]}{\delta\rho(\vec{r})} = v_{xc}[\rho(\vec{r})] \tag{2.23}$$

This means we have to start from some initial electronic density $(\rho_0(\vec{r}))$ and proceed to solve Eq. 2.21. We obtain a new density which is used to solved Eq. 2.21 until self-consistency is reached [6, 7, 11]. DFT does not provide a description of how to construct the exchange-correlation functional, $E_{xc}[\rho(\vec{r})]$, it only stipulates the existence of $E_{xc}[\rho(\vec{r})]$ as a universal functional of the density $\rho(\vec{r})$, which is valid for all systems. Several approximations have been constructed (for the accurate description of the exchange-correlation energy, $e_{xc}^{inif}[\rho(\vec{r})]$), such as the *local-density approximation* (LDA) [11]. This formulation is established in the original Kohn–Sham theory, and the idea is that a system with inhomogeneous charge distribution is treated as having a *locally-homogeneous* spatial uniform distribution of $\rho(\vec{r})$:

$$E_{xc}^{LDA}[\rho(\vec{r})] = \int \rho(\vec{r})e_{xc}^{unif}[\rho(\vec{r})]d^3\vec{r} \tag{2.24}$$

where an approximate analytical expression for $e_{xc}^{unif}[\rho(\vec{r})]$ comes from the e_x Hartree–Fock exchange term (Slater determinant), and e_c is fitted from quantum Monte Carlo parametrizations of homogeneous electron gases of varying densities [12]. LDA offers a good level of accuracy for highly-homogeneous systems, and even for realistic non-homogeneous systems; while one of the drawbacks is its overestimation of binding energies for molecules and solids [11, 13]. Some other high-level approximations to the $e_{xc}^{unif}[\rho(\vec{r})]$ have been made, such as the *generalized gradient approximation* (GGA) [11, 14, 15], for which the exchange-correlation energy (for a spin-unpolarized system) is expressed as:

$$E_{xc}^{GGA}[\rho(\vec{r})] = \int f[\rho(\vec{r}), \nabla\rho(\vec{r})]d^3\vec{r} \tag{2.25}$$

in which the density $\rho(\vec{r})$ is expanded in terms of the gradient (∇) operator. This approximation is valid for systems with slowly-varying densities.

In order to solve the Kohn–Sham equations, we need to select a suitable (sufficiently large) finite basis set (in theory one will need an infinite basis set for a precise description of the molecular orbital). We can express the wave-fuction ψ in terms of Gaussian-type-orbitals (GTOs), in either cartesian or polar coordinates:

$$\psi(\vec{r}) = Ax^{l_x}y^{l_y}z^{l_z}\exp(-\zeta r^2) \tag{2.26}$$

where A is a normalization constant, l determines the orbital, ζ is related to the width of the curve, and the r^2 gives the curve a Gaussian shape [16].

2.2.4 Functionals and Basis Sets

DFT calculations were performed using the **NWChem** (versions 4.7, 5.0 and 5.1) quantum chemistry package [17] and the Perdew–Wang exchange-correlation functional (PW91) [14, 15]. Test calculations using the Perdew–Burke–Ernzerhof (PBE) gradient-corrected exchange-correlation functional [18] produced qualitatively similar results while hybrid functionals (such as B3LYP) tend to underestimate atomization energies of d-metals, due to the inclusion of the Hartree–Fock exchange term, and its lack of a proper description of the nearly "free-electron" character of large metallic systems [19, 20].

Spherical Gaussian-type-orbital basis sets of double-ζ quality [21, 22] were used for Pd $(7s6p5d)/[5s3p2d]$, Ag $(7s6p5d)/[5s3p2d]$, Pt $(7s6p5d)/[6s3p2d]$ and Au $(7s5p5d)/[6s3p2d]$; combined with effective core potentials (ECP) in order to consider valence–electron only wavefunctions: 18 valence electrons were considered for Pd and Pt while 19 valence electrons were used for Ag and Au [23]. The ECP incorporates spin–orbit averaged relativistic for these four atoms.

Charge density fitting basis sets were used to speed up the evaluation of the Coulombic contributions [24]: Pd $(8s7p6d5f4g)/[8s6p6d3f2g]$, Ag $(9s4p5d3f4g)/[7s4p4d3f2g]$, Pt $(9s4p3d3f4g)/[9s4p3d3f2g]$, and Au $(9s4p4d3f4g)/[8s4p3d3f2g]$.

All calculations were performed spin unrestricted (geometry optimization for singlet spin states; after SCF optimization convergence a single point triplet spin state calculation is performed). Geometry optimizations were stopped when maximum force on atoms was less than 4×10^{-4} a.u. During the iterative process, a Gaussian smearing technique was adopted, with a smearing parameter of 0.136 eV for the fractional occupation of the one-electron energy levels [25]. Selected calculations were performed with basis sets of triple-ζ-plus-polarization quality to evaluate DFT mixing energies for comparison with Gupta values [21, 22].

For the fitting of the DFT-fit parameters for Pd–Au (see Chap. 7), DFT calculations on the solid phases were performed using the **PWscf** (Plane-Wave self-consistent-field) computational code [26], employing ultra-softpseudo-potentials (Dr. Giovanni Barcaro, CNR-Pisa). A total of 10 and 11 electrons were explicitly considered for Pd and Au, respectively. The PBE gradient-corrected exchange-correlation functional was used. Values of 40 Ry for the energy cutoff of the wave function and 160 Ry for the energy cutoff (1 Ry = 13.606 eV) of the electronic density have been shown to provide accurate results [as shown in previous work for the description of a monolayer phase of TiO_x on Pt(111)], and were thus employed [27]. All the calculations were performed by applying a smearing procedure of the energy levels with a Gaussian broadening of 0.002 Ry. The Brillouin zone was described by a $(10 \times 10 \times 10)$ grid.

2.3 Global Optimization Techniques

2.3.1 The Birmingham Cluster Genetic Algorithm

Genetic algorithms (GA) were developed by John Holland in the early 1970s, and since then, they have become successful tools in optimization problems in fields such as: chemistry, engineering and molecular modelling [28–30]. GAs belong to the class of stochastic optimization methods, which also includes techniques such as: evolutionary strategies, simulated annealing, and Monte Carlo optimization.

Structural global optimisation of a model potential function, consists of finding the configuration for which the PES is an absolute minimum: i.e. the GM. In recent years, GA techniques have been shown to be robust for finding the GM for a variety of types of nanoclusters [30, 31]. In a general way, genetic algorithms begin with an initial population of candidate solutions (in our studies, an initial population of starting cluster structures), which, through several iterative steps in the algorithm (crossover, mutation and natural selection), evolve towards the best solution of the problem (i.e the GM in the PES).

Our in-house Birmingham cluster genetic algorithm (BCGA) was written in order to find GM structures for several metallic and bimetallic systems, using the Gupta potential to model their interatomic interactions [30]. Its specific features and operators are defined as follows:

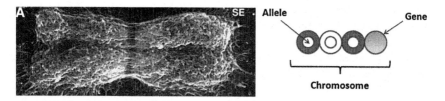

Fig. 2.2 *Left* field emission scanning electron microscopy image of a barley chromosome metaphase. *Right* schematic representation of the cluster information encoding in a string (*chromosome*) in the GA code [32]

Initial population the group of individuals which are going to be evolved by the genetic algorithm. In the BCGA, the initial population corresponds to a set of randomly generated clusters; where real valued Cartesian coordinates are generated within a cubic volume which is proportional to the number of atoms within the cluster. This initial population is then relaxed using a quasi-Newton (L-BFGS) minimization routine [33]. When we talk about individuals, we refer to a set of variables, known as *genes*. Genes form strings called *chromosomes*, which represent a trial solution of the problem. Individual *gene* values are know as *alleles*. Figure 2.2 shows a Field emission scanning electron microscopy image which illustrates a barley chromosome metaphase along with the simplified *chromosome* version implemented in the BCGA [32].

Fitness is defined (for a member of a certain population) as the quality of the trial solution represented by a *chromosome*, with respect to the function being optimized. In our GA optimizations, high values correspond to a better solution to the problem. In the BCGA, the total potential energy (V_i) of a cluster is first rescaled (Eq. 2.27), relative to the highest (V_{max}) and lowest (V_{min}) energy cluster in the current population:

$$\rho_i = \frac{V_i - V_{min}}{V_{max} - V_{min}} \tag{2.27}$$

A *fitness* function (F_i), Eq. 2.28, is used by the BGCA in order to determine, according to a probabilistic value (in which 0 is the worst option and 1 the best one) which individuals will survive from one generation to the next.

$$F_i = \frac{1}{2}[1 - \tanh(2\rho_i - 1)] \tag{2.28}$$

Selection of parents for crossover in the BCGA, individuals are selected to take part in the crossover based on their *fitness*. There are different methodologies, such as: (a) *roulette wheel selection*, in which a string is choosen at random, and selected for crossover if its *fitness* value is greater than a random number generated between 0 and 1; and (b) *tournament selection*, in which a number of *strings* are randomly selected from the population. The two highest *fitness strings* are then selected as parents from this tournament pool.

Fig. 2.3 Schematic representation of GA *crossover*

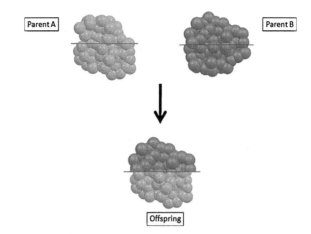

Crossover also know as "mating", is the exchange of genetic information between *chromosomes* (*string* parents). In the BCGA, parent structures are combined in order to generate *offspring*, by the cut-and-splice procedure of Deavon and Ho [30]. In our work, we have used the 1-point-weighted *crossover* (see Fig. 2.3), in which the cut position is based on the energies of the parents (the relative *fitness* of the parents). The BCGA slices the two clusters parent and combines complementary slices, with more atoms being chosen from the parent of highest fitness. In our implementation, only one offspring is generated.

Mutation as in nature, mutations help to avoid stagnation of the population (i.e. they increase population diversity). Mutations results in slight changes in the genetic information encoded in the *chromosome*: in other words, new genetic material is introduced into the *population*. In global optimization routines, introducing a mutation operator can help to avoid convergence to a non-optimal solution (a high-energy nanocluster structure). In the BCGA, a number of different mutation schemes have been encoded:

(a) *Replacement* one cluster is removed from the population and replaced by another generated at random.
(b) *Rotation* a rotation of the atomic coordinates is performed of the top half of the cluster relative to the bottom half, by a random angle.
(c) *Exchange* applied to bimetallic clusters, in which approximately one-third of the A type atoms in the cluster are exchanged for B atoms, without altering the original coordinates in the cluster
(d) *Displacement* approximately one-third of the atoms in the clusters are displaced to random positions. The selected atoms are also choosen at random.

Selection evolutionary principles are best applied at this stage during GA searches (i.e. the Darwinian principle of survival of the "fittest"). There are modifications which can be taken into account, such as accepting all mutant structures (or the contrary case), not accepting parents for the next generation, or always keeping among the population the individuals with higher *fitness* (i.e. the elitist strategy).

Optimization all processes, such as crossover, mutation and selection is repeated for a specified number of generations (in this work, number of generations = 400) until the population is considered to be converged (when the range of cluster energies in the population has not changed for a prescribed number of generations).

The *BCGA parameters* adopted in this work were: population size = 40 clusters; crossover rate = 0.8 (i.e. 32 offspring are produced per generation); crossover type = 1-point weighted; selection = roulette; mutation rate = 0.1; mutation type = mutate-move; number of generations = 400, and number of GA runs for each composition = 100 (for statistical purposes, as well as for large clusters, a large number of global optimization searches are needed due to the complexity of the PES of the nanoparticle). Most of the parameters are selected as default by the BCGA code (such as crossover and mutation rate). A large initial population and a vast number of generations are needed for a full exploration of the PES of these medium size clusters. BCGA global optimizations were stopped when there is no change in the population over 10 different generations (term = 10).

2.3.2 The Basin-Hopping Monte Carlo Algorithm

A more detailed homotop search was performed using a modified version of the Basin Hopping Monte–Carlo algorithm (BHMC) [34], which allows only "atom-exchange" for a fixed composition and structural configuration (see Fig. 2.4).

During our local BHMC optimizations, we carried out approximately 3,000 Monte–Carlo steps, with a thermal energy $k_B T$ of 0.02 eV. This low value is appropriate for performing a localized search of deep regions of the chosen structural funnel in the PES. We found that in some cases, the GA approach was not able to identify the most favorable chemical ordering, whereas this was easily found using the BHMC algorithm. This is in line with recent observations that the optimal strategy in configurational searches is to take initially only structural moves, and subsequently to refine the lowest-energy structures via atom-exchange moves [35, 36 and Ferrando and Rossi (2007, personal communication)].

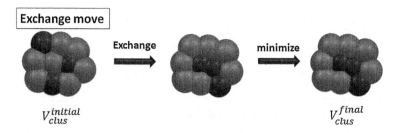

Fig. 2.4 Schematic representation of the "atom-exchange" local relaxation, implemented in the BHMC algorithm

Fig. 2.5 98-atom Leary
Tetrahedron (LT) structure
with T_d symmetry. Each
colour represents a different
atomic shell

2.3.3 Shell Optimization Routine

For high-symmetry polyhedral cluster geometries, a substantial reduction in
the search space is obtained if all sets of symmetry-equivalent atoms, which we
have termed *atomic shells*, in a particular structure are constrained to be of the
same chemical species [37]. It should be noted that these *atomic shells* are, in a
more rigorous group theoretical sense, known as *orbits* of the point group [1]. For a
given geometrical structure, this reduces the number of inequivalent compositional
and permutational isomers (homotops) to 2^S, where S is the number of atomic shells.

We have used our ***shell optimization routine*** for the highly symmetric 98-atom
Leary tetrahedron (LT) structures (see Chaps. 4, 7), with ideal T_d symmetry. $S = 9$
shells and, in order of increasing distance from the centre of the cluster, these
shells have 4:12:12:12:4:6:12:12:24 atoms, resulting in $2^9 = 512$ T_d symmetry LT
isomers. The various shells of the LT structure are indicated in Fig. 2.5 by different colours. Using the *shell-optimization routine*, it is possible to conduct a
systematic investigation of all high-symmetry chemical arrangements for a given
structural motif, with greatly reduced computational effort.

2.3.4 Combined EP-DFT Approach

In this work, we have followed a *combined approach* (see Fig. 2.6), in which initial
global optimizations are carried out using our GA (coupled with the Gupta
potential) and once we have found a putative global minimum (GM), we perform an
"atom-exchange" routine (a modified Basin Hopping Monte Carlo, in collaboration
with Professor Alessandro Fortunelli and Dr. Giovanni Barcaro, CNR-Pisa).

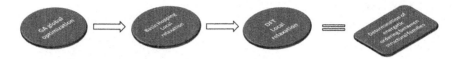

Fig. 2.6 Schematic representation of our *combined approach*

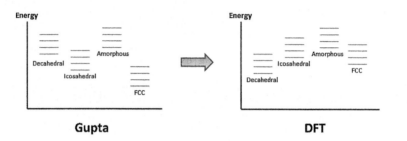

Fig. 2.7 *Left* putative GM *structural families* found at the Gupta level of theory. The energy orderings are modified after performing high-level DFT calculations *right*

This allows us to corroborate, at the Gupta level, the *chemical ordering* of the structure (i.e. the way in which atoms segregate to cluster surface sites). Once we have located a variety of putative GM structures (as well as high-energy isomers) for different bimetallic systems, we proceed to construct a database of structures (varying from crystalline fcc-type structures, to decahedral, icosahedral and amorphous type, *structural families*); and we place them in competition at the high level of theory, by carrying out DFT local-relaxations on these structures (see Fig. 2.7) [36, 38, 39].

In contrast to the search of the empirical potential (EP) potential energy surface using the GA method, we cannot guarantee that our combined EP/DFT approach will be equally successful in the search of the GM at the DFT level. For this reason, the expressions "putative GM" or "lowest-energy structure" are used when discussing the results of DFT calculations.

2.4 Energetic Analysis

In order to analyse cluster stability, some energetic quantities need to be defined. At the Gupta potential level of theory, the *binding energy* per atom of a cluster (E_b^{Gupta}), can be calculated as:

$$E_b^{\text{Gupta}} = -\frac{V_{\text{clus}}}{N} \qquad (2.29)$$

where V_{clus} is the total potential energy of the cluster (pure metal or bimetallic) and N is the total number of atoms in the cluster. In our work, positive values of E_b^{Gupta} indicate stable cluster atomic configurations (i.e. lower E_b^{Gupta} values would indicate less favourable atomic arrangements). Another criterion for determining relative stability (both at the Gupta and DFT level) is the *second difference in binding energy*, $\Delta_2 E_b(N)$. This quantity is used in pure clusters in order to compare the relative stability of a cluster of size N, with respect to its neighbours:

$$\Delta_2 E_b(N) = 2E_b(N) - E_b(N-1) - E_b(N+1) \tag{2.30}$$

In the case of mixed clusters, Eq. 2.30 can be defined as follows:

$$\Delta_2 E_b(A_m B_n) = E_b(A_{m+1} B_{n-1}) + E_b(A_{m-1} B_{n+1}) - 2E_b(A_m B_n) \tag{2.31}$$

where peaks in both $\Delta_2 E_b(N)$ and $\Delta_2 E_b(A_m B_n)$ often coincide with discontinuities in the mass spectra [40].

Another useful quantity at the Gupta level is the *excess energy* (i.e. the *mixing energy*) Δ_N^{Gupta} defined for clusters of fixed size but different composition [41]:

$$\Delta_N^{Gupta} = E_N^{Gupta}(A_m B_n) - m\frac{E_N^{Gupta}(A_N)}{N} - n\frac{E_N^{Gupta}(B_N)}{N} \tag{2.32}$$

where $E_N^{Gupta}(A_m B_n)$ represents the total energy of a given cluster (e.g. all possible compositions for N-atom: Pd–Pt, Ag–Pt, Pd–Au and Ag–Au clusters) calculated at the Gupta level and $E_N^{Gupta}(A_N)$ and $E_N^{Gupta}(B_N)$ represent the total energies of the GM of the pure metal clusters (i.e. Pt_N, Ag_N, Pd_N and Au_N). Δ_N^{Gupta} quantifies the degree of mixing (the energy associated with alloying) between the two different metals. The most negative values of Δ_N^{Gupta} represent those compositions at which mixing is most favourable, and thus, the more stable clusters.

In order to analyse trends in chemical order as a function of size and composition (See Chap. 7), it is convenient to define an order parameter with the following characteristics: positive when phase separation (segregation) takes place, close to zero when disordered mixing occurs, and negative when mixing and layer-like structure co-exist. The chemical order parameter σ is defined as:

$$\sigma = \frac{N_{Pd-Pd} + N_{Au-Au} - N_{Pd-Au}}{N_{Pd-Pd} + N_{Au-Au} + N_{Pd-Au}} \tag{2.33}$$

where N_{ij} (with i, j = Pd, Au) is the number of nearest neighbour $i - j$ bonds. An order parameter of this type has proven to be useful in the description of short range order in binary bulk alloys and surfaces [42].

The *DFT binding energies* (E_b^{DFT}) for N-atom pure metal clusters can be calculated as the difference (per atom) between the total electronic energy of the cluster ($E_{total}^{DFT}(A_N)$) and N times the energy of a single atom $E_{atom}^{DFT}(A)$:

$$E_b^{\mathrm{DFT}} = -\frac{1}{N}\left[E_{\mathrm{total}}^{\mathrm{DFT}}(A_N) - N \cdot E_{\mathrm{total}}^{\mathrm{DFT}}(A)\right] \tag{2.34}$$

For bimetallic clusters, E_b^{DFT} is calculated using a similar expression:

$$E_b^{\mathrm{DFT}} = -\frac{1}{N}\left[E_{\mathrm{total}}^{\mathrm{DFT}}(A_m B_n) - m \cdot E_{\mathrm{atom}}^{\mathrm{DFT}}(A) - n \cdot E_{\mathrm{atom}}^{\mathrm{DFT}}(B)\right] \tag{2.35}$$

The excess energy Δ_N^{DFT}, can also be defined at the DFT level as follows:

$$\Delta_N^{\mathrm{DFT}} = E_N^{\mathrm{DFT}}(A_m B_n) - m E_{A_N}^{\mathrm{DFT}} - n E_{B_N}^{\mathrm{DFT}} \tag{2.36}$$

where $E_N^{\mathrm{DFT}}(A_m B_n)$ is the total electronic energy of the cluster, with size N and composition $(A_m B_n)$, and $E_{A_N}^{\mathrm{DFT}}$ and $E_{B_N}^{\mathrm{DFT}}$ are the total electronic energies of the GM of the pure metal clusters A_N and B_N, respectively.

References

1. D.J. Wales, *Energy Landscapes, with Applications to Clusters, Biomolecules and Glaseses* (Cambridge University Press, Cambridge, 2003)
2. R.P. Gupta, Phys. Rev. B **23**, 6265 (1981)
3. F. Cleri, V. Rosato, Phys. Rev. B **48**, 22 1993
4. L.H. Thomas, Proc. Camb. Phil. Soc. **23**, 542 (1927)
5. E. Fermi, Z. Phys. **48**, 73 (1928)
6. R.G. Parr, W. Yang, *Density-Functional Theory of Atoms and Molecules* (Oxford University Press, Oxford, 1989)
7. M. Finnis, *Interatomic Forces in Condensed Matter* (Oxford University Press, Oxford, 2003)
8. P.W. Atkins, R.S. Friedman, *Molecular Quantum Mechanics*, 3rd edn. (Oxford University Press, Oxford, 1997)
9. P. Hohenberg, W. Kohn, Phys. Rev. **136**(3B), B864
10. W. Kohn, L.S. Sham, Phys. Rev. **140**(4A), A1133 (1965)
11. W. Kohn, Rev. Mod. Phys. **71**(5), 1253 (1999)
12. D.M. Ceperley, B.J. Alder, Phys. Rev. Lett. **45**, 566 (1980)
13. A. Khein, D.J. Singh, C.J. Umrigar, Phys. Rev. B **51**, 4105 (1995)
14. J.P. Perdew, Y. Wang, Phys. Rev. B **33**, 8800 (1986)
15. J.P. Perdew, J.A. Chevary, S.H. Vosko, K.A. Jackson, M.R. Pederson, D.J. Singh, C. Fiolhaus, Phys. Rev. B **46**, 6671 (1992)
16. F. Jensen, *Introduction to Computational Chemistry* (Springer, Berlin, 1999)
17. R.A. Kendall, E. Aprà, D.E. Bernholdt, E.J. Bylaska, M. Dupuis, G.I. Fann, R.J.Harrison, J. Ju, J.A. Nichols, J. Nieplocha, T.P. Straatsma, T.L. Windus, A.T. Wong, Comput. Phys. Commun. **128**, 260 (2000)
18. J.P. Perdew, K. Burke, M. Ernzerhof, Phys. Rev. Lett. **77**, 3865 (1996)
19. E. Aprà, A. Fortunelli, J. Mol. Struct. (Theochem) **501**, 251 (2000)
20. J. Paier, M. Marsman, G. Kresse, J. Chem. Phys. **127**, 024103 (2007)
21. A. Schäfer, C. Huber, R. Ahlrichs, J. Chem. Phys. **100**, 5829 (1994)
22. See: ftp://ftp.chemie.uni-karlsruhe.de/pub/basen/
23. D. Andrae, U. Haeussermann, M. Dolg, H. Stoll, H. Preuss, Theor. Chim. Acta. **77**, 123 (1990)
24. F. Weigend, M. Haser, H. Patzel, R. Ahlrichs, Chem. Phys. Lett. **294**, 143 (1998)
25. E. Aprà, A. Fortunelli, J. Phys. Chem. A **107**, 2934 (2003)

26. S. Baroni, A. Del Corso, S. de Gironcoli, P. Giannozzi, http://www.pwscf.org
27. G. Barcaro, F. Sedona, A. Fortunelli, G. Granozzi, J. Phys. Chem. C **111**, 6095 (2007)
28. D. Lawrence, *Handbook of Genetic Algorithms*. ed. Van Nostrand Reinhold, New York (1991)
29. J. Holland, *Adaptation in Natural and Artificial Systems*. Springer, Berlin (1975)
30. R.L. Johnston, Dalton Trans. 4193 (2003)
31. B. Hartke, *Applications of Evolutionary Computation in Chemistry*, in R.L. Johnston, (Springer, Berlin, 2004)
32. E. Schroeder-Reiter, F. Pérez-Willard, U. Zeile, G. Wanner, J. Struct. Bio. **165**, 97 (2009)
33. R.H. Byrd, P. Lu, J. Nocedal, Zhu C, J. Scient. Comp. **5**, 1190 (1995)
34. J.P.K. Doye, D.J. Wales, J. Phys. Chem. A **101**, 5111 (1997)
35. L.O. Paz-Borbón, T.V. Mortimer-Jones, R.L. Johnston, A. Posada-Amarillas, G. Barcaro, A. Fortunelli, Phys. Chem. Chem. Phys. **9**, 5202 (2007)
36. R. Ferrando, R.L. Johnston, A. Fortunelli, Phys. Chem. Chem. Phys. **10**, 640 (2008)
37. N.T. Wilson, R.L. Johnston, J. Mater. Chem. **12**, 2913 (2002)
38. L.O. Paz-Borbón, R.L. Johnston, G. Barcaro, A. Fortunelli, J. Chem. Phys. **128**, 134517 (2008)
39. L.O. Paz-Borbón, R.L. Johnston, G. Barcaro, A. Fortunelli, J. Phys. Chem. C **111**, 2936 (2007)
40. W.D. Knight, K. Clememger, W.A. de Heer, W.A. Saunders, M.Y. Chou, M.L. Cohen, Phys. Rev. Lett. **52**, 2141 (1984)
41. R. Ferrando, R.L. Johnston, J. Jellinek, Chem. Rev. **108**, 845 (2008)
42. F. Aguilera-Granja, A. Vega, J. Rogan, X. Andrade, G. García, Phys. Rev. B. **74**, 224405 (2006)

Chapter 3
34-Atom Pd–Pt Clusters

3.1 Introduction

Computational studies, based on the Gupta many-body potential and a genetic algorithm search method, have given new insights into the geometries and segregation properties of Pd–Pt nanoalloys with up to 56 atoms [1–3]. Using average parameters for the mixed (Pd–Pt) interactions [1] that were derived based on considerations of the mixing behavior of the bulk Pd–Pt alloys, the experimentally observed tendency of Pt and Pd atoms to preferentially occupy interior (core) and exterior (surface) sites, respectively, has been confirmed by computational simulations. Different types of structures (e.g. icosahedral, cubic close-packed, decahedral, or amorphous geometries) adopted by the lowest energy isomers for Pd, Pt, and Pd–Pt clusters were also identified. Recent calculations by Rossi et al. [4] based on the Gupta potential showed that pure 34-atom Pd and Pt clusters are incomplete decahedra, and in the case of mixed clusters they exhibit the same structural motif for several different compositions. Calculations by Fernandez et al. [5] also showed that the $Pd_{shell}Pt_{core}$ segregation predicted by the Gupta potential simulations are reproduced at higher levels of theory, although the energy ordering of the permutational isomers may differ from Gupta to DFT calculations.

3.2 Computational Details

The GA parameters adopted for the study of 34-atom Pd–Pt clusters were: population size = 40 clusters; crossover rate = 0.8; crossover type = 1-point weighted; selection = roulette; mutation rate = 0.1, mutation type = mutate-move; number of generations = 400; number of GA runs for each composition = 50. For the specific composition $Pd_{24}Pt_{10}$, 1,000 GA runs were performed in order to study the distribution of several structural arrangements, due to the high

L. O. Paz Borbón, *Computational Studies of Transition Metal Nanoalloys*,
Springer Theses, DOI: 10.1007/978-3-642-18012-5_3,
© Springer-Verlag Berlin Heidelberg 2011

number of homotops. A Basin Hopping Monte Carlo algorithm [6] was also used to carried out approximately 3,000 Monte Carlo steps, with a thermal energy $k_B T$ of 0.02 eV. DFT calculations were performed using double-ζ quality basis set (see Chap. 2).

3.3 Results and Discussion

3.3.1 Gupta Potential Calculations on Pd_mPt_{34-m}

Figure 3.1a shows a plot of Δ_{34}^{Gupta} against Pd concentration (m), for the Gupta potential GM for all compositions Pd_mPt_{34-m}. The lowest values of Δ_{34}^{Gupta} were found for the compositions $Pd_{21}Pt_{13}$ and $Pd_{24}Pt_{10}$, thus indicating that these are relatively stable GM. This is confirmed in Fig. 3.1b, which shows the corresponding $\Delta_2 E$ plot, with peaks at $m = 21$, 24, and 30. This is because of the high-symmetry core–shell segregated nature of these GM, which exhibit polyhedral or

Fig. 3.1 Excess energy **a** and second difference in energy **b** for 34-atom Pd–Pt clusters, modelled by the Gupta potential

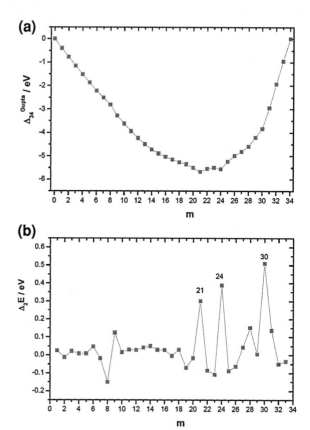

polygonal cores of Pt (a centered icosahedron in $Pd_{21}Pt_{13}$, a tetrahedron in $Pd_{24}Pt_{10}$, and a rectangle in $Pd_{30}Pt_4$). According to the Δ_{34}^{Gupta} and $\Delta_2 E$ analysis (see Fig. 3.1), one should expect to find relatively stable Pd–Pt structures in the composition range $Pd_m Pt_{34-m}$ ($m \approx 17$–28).

Based on these preliminary calculations, a detailed study of clusters in the range $Pd_{17}Pt_{17}$ to $Pd_{28}Pt_6$ was undertaken. A number of low-energy structural families (motifs) were identified. These structures are shown in Fig. 3.2 for the particular case of the 24–10 composition. Figure 3.2a shows a mixed decahedral (Dh)/close-packed (cp) motif in which a tetrahedral (T) core of 10 atoms is surrounded by 12 atoms growing on the (111) faces and 12 atoms growing along the edges of the tetrahedron. Figure 3.2b shows a mixed Dh/cp motif, presenting a 14-atom core with the structure of a trigonal bipyramid (a "double tetrahedron", DT) surrounded by 18 atoms growing on the (111) faces and 2 atoms growing along one of the edges of the double tetrahedron. Figure 3.2c shows an incomplete, somewhat distorted 38-atom truncated octahedron (TO), lacking 4 atoms of a (100) face.

Figure 3.2e shows an incomplete Marks decahedron [7]. Finally, Fig. 3.2d, f and g show three different polyicosahedral (pIh) structures [8], of which Fig. 3.2d is a low-symmetry configuration, Fig. 3.2g is an incomplete and slightly distorted 38-atom 6-fold pIh^6 pancake, lacking two dimers on the basal ring, and Fig. 3.2f is an incomplete 45-atom 5-fold pIh^5 (i.e. an incomplete anti-Mackay icosahedron).

In Fig. 3.3, we plot the variation of Δ_{34}^{Gupta} with composition for the lowest energy homotop of each structural motif of Fig. 3.2 with the exception of the low-symmetry pIh, which being a disordered structure is sometimes difficult to locate. As a general trend, decahedral motifs dominate for 34-atom Pd–Pt clusters (see Tables 3.1, 3.2 and 3.3). Many incomplete Marks Dh motifs are found as global

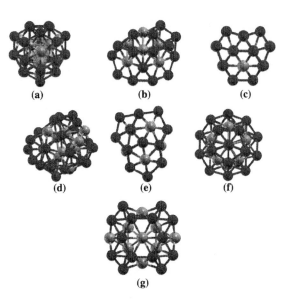

Fig. 3.2 Different structural motifs found for the composition $Pd_{24}Pt_{10}$. Pd and Pt atoms are indicated by *dark* and *light color spheres*, respectively. **a** Dh–cp(T), **b** Dh–cp(DT), **c** TO, **d** low-symmetry pIh, **e** Marks Dh, **f** incomplete 5-fold pIh, **g** incomplete 6-fold pIh

(a)   (b)   (c)

(d)   (e)   (f)

(g)

Fig. 3.3 Plot of excess energy $\Delta_{34}^{\text{Gupta}}$ for different structural motifs at compositions Pd_mPt_{34-m} varying from $Pd_{17}Pt_{17}$ to $Pd_{28}Pt_6$

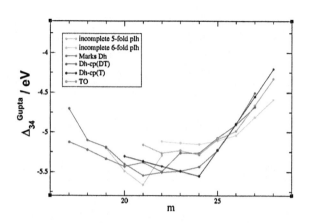

Table 3.1 Comparison of GM (global minimum) structures for 34-atom Pd–Pt clusters for the Gupta potential and DFT calculations, in the $m = 17$–30 composition range (including pure Pd_{34} and Pt_{34} clusters)

Structure	GM(EP)	Symm	$E_{\text{Total}}^{\text{Gupta}}/eV$	GM(DFT)	Symm	$E_{\text{Total}}^{\text{DFT}}/(au)$
Pd_{34}	Incomplete decahedral	C_2	−116.9679	Incomplete decahedral	C_2	−4352.4449
$Pd_{30}Pt_4$	Incomplete 6-fold pIh	D_{2h}	−127.9682	Incomplete 6-fold pIh	D_{2h}	−4318.3975
$Pd_{29}Pt_5$	Incomplete 6-fold pIh	C_{2v}	−130.1384	Incomplete 6-fold pIh	C_{2v}	−4309.8891
$Pd_{28}Pt_6$	Incomplete 6-fold pIh	D_{2h}	−132.3026	Dh–cp(DT)	C_s	−4301.4036
$Pd_{27}Pt_7$	Incomplete 6-fold pIh	C_{2v}	−134.3152	Dh–cp(DT)	C_s	−4292.8988
$Pd_{26}Pt_8$	TO	C_{2v}	−136.2842	Dh–cp(DT)	C_s	−4284.3918
$Pd_{25}Pt_9$	Dh–cp(T)	C_{3v}	−138.3158	Dh–cp(DT)	C_s	−4275.8822
$Pd_{24}Pt_{10}$	Dh–cp(T)	T_d	−140.4348	Dh–cp(DT)	C_s	−4267.3691
$Pd_{23}Pt_{11}$	Dh–cp(T)	C_s	−142.1643	Dh–cp(DT)	C_s	−4258.8511
$Pd_{22}Pt_{12}$	Dh–cp(DT)	C_s	−144.0010	Dh–cp(DT)	C_2	−4250.3378
$Pd_{21}Pt_{13}$	Incomplete 5-fold pIh	C_{5v}	−145.9230	Dh–cp(DT)	C_s	−4241.8184
$Pd_{20}Pt_{14}$	Incomplete 5-fold pIh	C_s	−147.5432	Dh–cp(DT)	C_s	−4233.3002
$Pd_{19}Pt_{15}$	Marks Dh	C_1	−149.1797	Dh–cp(DT)	C_s	−4224.7849
$Pd_{18}Pt_{16}$	Low-symmetry pIh	C_s	−150.8852	Dh–cp(DT)	C_s	−4216.2646
$Pd_{17}Pt_{17}$	Marks Dh	C_1	−152.5605	Dh–cp(DT)	C_s	−4207.7480
Pt_{34}	Incomplete decahedral	C_2	−177.8761	Incomplete decahedral	C_2	−4062.8905

Total DFT energies ($E_{\text{Total}}^{\text{DFT}}$) are given in atomic units (au), while Gupta potential energies ($E_{\text{Total}}^{\text{Gupta}}$) are given in electron-volts (eV)

Table 3.2 Structures, symmetries and total energies (Gupta level) for specific 34-atom Pd–Pt clusters in the $m = 31$–33, and $m = 1$–16 composition range

Composition	Motif	Symmetry	$E_{\text{Total}}^{\text{Gupta}}/eV$
$Pd_{33}Pt_1$	Marks Dh	C_s	-119.702171
$Pd_{32}Pt_2$	Low-symmetry pIh	C_1	-122.470613
$Pd_{31}Pt_3$	Low-symmetry pIh	C_1	-125.288139
$Pd_{16}Pt_{18}$	Low-symmetry pIh	C_1	-154.101708
$Pd_{15}Pt_{19}$	Marks Dh	C_s	-155.890715
$Pd_{14}Pt_{20}$	Marks Dh	C_1	-157.511383
$Pd_{13}Pt_{21}$	Low-symmetry pIh	C_1	-158.722074
$Pd_{12}Pt_{22}$	Marks Dh	C_1	-160.610847
$Pd_{11}Pt_{23}$	Marks Dh	C_s	-162.111451
$Pd_{10}Pt_{24}$	Marks Dh	C_1	-163.580682
Pd_9Pt_{25}	Marks Dh	C_1	-165.033719
Pd_8Pt_{26}	Marks Dh	C_1	-166.362200
Pd_7Pt_{27}	Marks Dh	C_1	-167.839772
Pd_6Pt_{28}	Marks Dh	C_1	-169.335320
Pd_5Pt_{29}	Marks Dh	C_1	-170.783213
Pd_4Pt_{30}	Marks Dh	C_1	-172.222597
Pd_3Pt_{31}	Marks Dh	C_1	-173.652960
Pd_2Pt_{32}	Marks Dh	C_1	-175.061020
Pd_1Pt_{33}	Incomplete decahedral	C_1	-176.481095

minima in the Pt_{rich} composition range. At the Gupta level, the lowest excess energy structure occurs for $Pd_{21}Pt_{13}$, which has a complete icosahedral Pt_{13} core. It is followed in energy by $Pd_{24}Pt_{10}$, which has a tetrahedral Pt_{10} core. Higher excess energy structures correspond to incomplete 6-fold pancakes as well as incomplete TO structures. The main conclusion of this analysis is that at the empirical potential level one finds a complex crossover between several different structural families that are rather close in energy: the energy difference between the GM and the first high-energy isomer ranges from essentially zero to about 0.2 eV in the chosen composition interval.

3.4 DFT Calculations

First of all, for each composition the most stable homotop at the Gupta level was locally optimized at the DFT level. Figure 3.4 shows a plot (red line) of the Δ_{34}^{DFT} calculated values. The jagged nature of this plot makes it unlikely that these structures correspond to GM on the DFT energy hypersurface. It is interesting to note that the lowest Δ_{34}^{DFT} value on this plot is the Dh–cp(DT) structure for $Pd_{22}Pt_{12}$, which was only obtained as the GM for this composition at the Gupta level. Because of its high stability, the Dh–cp(DT) structure was then locally optimized at the DFT level for the selected $Pd_{17}Pt_{17}$ to $Pd_{28}Pt_6$ composition range. The Dh–cp(DT) structures were constructed for each composition, with their

Table 3.3 Ordering of structural isomers of 34-atom Pd–Pt clusters as a function of composition, for Gupta potential and DFT calculations

Composition	Predicted EP motifs	Predicted DFT motifs
$Pd_{17}Pt_{17}$	Marks Dh < low-symmetry pIh	Dh–cp(DT) < Marks Dh < low-symmetry pIh
$Pd_{18}Pt_{16}$	Low-symmetry pIh < Marks Dh	Dh–cp(DT) < low-symmetry pIh
$Pd_{19}Pt_{15}$	Marks Dh < Dh–cp(DT) < TO < low-symmetry pIh	Dh–cp(DT) < Marks Dh
$Pd_{20}Pt_{14}$	Incomplete 5-fold pIh < Dh–cp(DT) < TO < Marks Dh < Dh–cp(T) < low-symmetry pIh	Dh–cp(DT) < incomplete 5-fold pIh
$Pd_{21}Pt_{13}$	Incomplete 5-fold pIh < TO < Marks Dh < Dh–cp(T) < low-symmetry pIh	Dh–cp(DT) < TO < incomplete 5-fold pIh
$Pd_{22}Pt_{12}$	Dh–cp(DT) < Marks Dh < TO < Dh–cp(T)	Dh–cp(DT)
$Pd_{23}Pt_{11}$	Dh–cp(T) < Dh–cp(DT) < Marks Dh < TO < low-symmetry pIh	Dh–cp(DT) < TO < Dh–cp(T)
$Pd_{24}Pt_{10}$	Dh–cp(T) < Dh–cp(DT) < TO < Marks Dh < low-symmetry pIh	Dh–cp(DT) < TO < low-symmetry pIh < Dh–cp(T)
$Pd_{25}Pt_{9}$	Dh–cp(T) < TO < Marks Dh < Dh–cp(DT)	Dh–cp(DT) < Dh–cp(T) < TO
$Pd_{26}Pt_{8}$	Incomplete 6-fold pIh < TO < Marks Dh < Dh–cp(DT) < Dh–cp(T)	Dh–cp(DT) < TO
$Pd_{27}Pt_{7}$	Incomplete 6-fold pIh < Marks Dh < TO < Dh–cp(T) < Dh–cp(DT)	Dh–cp(DT) < incomplete 6-fold pIh
$Pd_{28}Pt_{6}$	Incomplete 6-fold pIh < TO < Dh–cp(T)	Dh–cp(DT) < incomplete 6-fold pIh
$Pd_{29}Pt_{5}$	Incomplete 6-fold pIh < incomplete distorted 5-fold pIh < low-symmetry pIh	Incomplete 6-fold pIh
$Pd_{30}Pt_{4}$	Incomplete 6-fold pIh	Incomplete 6-fold pIh

corresponding chemical ordering optimized by the BHMC "exchange-only" algorithm [6]. The blue line in Fig. 3.4 is a plot of Δ_{34}^{DFT} values after DFT reoptimization of Dh–cp(DT) structures across this composition range. The small fluctuations on Δ_{34}^{DFT} for the Dh–cp(DT) structures may be due to the lowest DFT homotops not being found. The following analysis will show that this structure is the putative GM at the DFT level for all compositions within the $Pd_{17}Pt_{17}$-$Pd_{28}Pt_{6}$ range. A detailed study now will be presented for two particular compositions: $Pd_{24}Pt_{10}$ and $Pd_{17}Pt_{17}$ for which we conducted DFT local geometry optimization for structures representative of several structural families; in a similar way as in previous work by Barcaro et al. on Ag–Cu and Au–Cu clusters [9].

3.4.1 Composition $Pd_{24}Pt_{10}$

At the Gupta level, the GM at composition $Pd_{24}Pt_{10}$ corresponds to a highly symmetrical structure: a tetrahedral core of 10 Pt atoms surrounded by the remaining 24 Pd atoms (Fig. 3.2a). An exhaustive search (1,000 GA runs) was then performed at

Fig. 3.4 Δ_{34}^{DFT} curve obtained by local geometry optimizations of the global minima predicted at the Gupta potential level (*red dashed line*) and of Dh–cp(DT) structures (*blue line*)

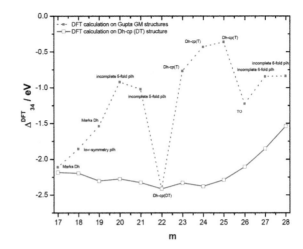

Table 3.4 Energy ordering predicted by the Gupta potential and DFT calculations, at the $Pd_{24}Pt_{10}$ composition

Motif	Symmetry	$\Delta E_{34}^{Gupta}/eV$	$\Delta E_{34}^{DFT}/eV$
Dh–cp(DT)	C_s	0.115	0.0000
TO	C_s	0.271	0.6911
Marks Dh	C_1	0.285	0.7537
Incomplete 6-fold pIh	D_{4h}	0.401	1.2571
Low symmetry pIh	C_1	0.427	1.5619
Dh–cp(T)	T_d	0.000	1.9584
Incomplete 5-fold pIh	C_{5h}	1.118	2.0599

this composition, using the BCGA code. This search gave us a wide distribution of several structural arrangements. Following a previously proposed protocol [9], each of the lowest energy homotops for each structural motif was subjected to DFT local optimization. Figure 3.2 shows the lowest energy homotops found for each of the seven low-energy structural arrangements (at the Gupta level) for this composition.

Table 3.4 shows the energetic ordering and relative energies of the different isomers as well as their corresponding symmetries. It is apparent that, after performing the DFT local optimization a substantial change in the energetic ordering occurs. The highly symmetric Dh–cp(T) structure, which is the GM at the Gupta level, is found to be one of the motifs with the highest Δ_{34}^{DFT} value (see Fig. 3.5). In agreement with the results from the previous subsection, the new putative GM corresponds to the Dh–cp(DT) structure. We have performed a similar (though not as complete) analysis at other compositions by locally optimizing various higher-energy isomers. In the chosen composition interval the Dh–cp(DT) structure turned out to have the lowest-energy in all cases.

In principle, following the complete protocol [9], one should locally optimize several low energy structures and homotops for each structural family, as it is found by Fernandez and co-workers. They found that DFT relaxation can also

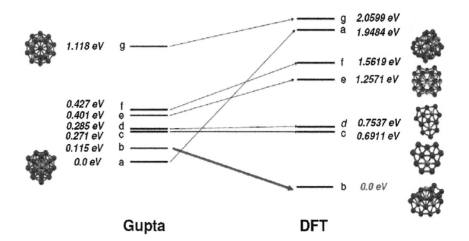

Fig. 3.5 Structures of seven low energy 34-atom structural motifs commonly found for Pd–Pt clusters. The structures, found as the lowest energy homotops using the Gupta potential for Pd$_{24}$Pt$_{10}$, are: **a** Dh–cp(T); **b** Dh–cp(DT); **c** TO; **d** Marks Dh; **e** incomplete 6-fold pIh; **f** low-symmetry pIh; **g** incomplete 5-fold pIh. The figure shows how the relative stabilities of these structures change on reoptimising the Gupta potential minima at the DFT level. The numbers are the energies of the structural motifs, in eV, relative to the lowest energy structure, in each case. The *thick arrow* shows the relative stabilisation of the Dh–cp(DT) motif at the DFT level. *Note* Gupta energies are shifted compared to the corresponding DFT ones, as we are comparing two different levels of theory

change the homotop energy ordering within each family [5]. However, this effect is usually less dramatic when comparing different structural motifs. This was confirmed by DFT calculations on several higher-energy configurations from selected structural families.

3.4.2 Segregation Effects for Pd$_{17}$Pt$_{17}$ Clusters

As mentioned previously, segregation of Pd atoms to the surface and Pt atoms to the core seems to be a general tendency for Pd–Pt clusters [1–3, 10–12]. This is interesting as Pd–Pt clusters are among the few systems in which the lighter atoms tend to segregate to the surface. As Pd$_{17}$Pt$_{17}$ has a 50:50 composition of Pd and Pt atoms, it represents a good candidate to study surface segregation effects.

Several structures were generated using the Gupta potential and the GA search technique. It was found that Marks Dh type motifs are preferred over pIh structures at the Gupta potential level, with the Dh–cp(DT) structure somewhat higher in energy. The lowest-energy homotops corresponding to these three structural families were then selected and the Pd and Pt atom positions exchanged before performing local geometry optimization at the DFT level. In Fig. 3.6 the three lowest-energy motifs for the Pd$_{17}$Pt$_{17}$ cluster, the corresponding inverted structures, and the relative energies are shown. It can be seen that swapping Pd and Pt atoms

Fig. 3.6 Structural motifs at the $Pd_{17}Pt_{17}$ composition. The *left column* shows low energy structures predicted by the Gupta potential and reoptimized at the DFT level. The *right column* shows the DFT reoptimized inverted structures in which Pd (*dark color*) and Pt (*light color*) atoms have been swapped

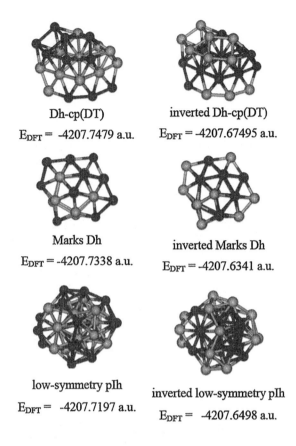

Dh-cp(DT)

$E_{DFT} = $ -4207.7479 a.u.

inverted Dh-cp(DT)

$E_{DFT} = $ -4207.67495 a.u.

Marks Dh

$E_{DFT} = $ -4207.7338 a.u.

inverted Marks Dh

$E_{DFT} = $ -4207.6341 a.u.

low-symmetry pIh

$E_{DFT} = $ -4207.7197 a.u.

inverted low-symmetry pIh

$E_{DFT} = $ -4207.6498 a.u.

always increases the total energy of the system, thus confirming that Pd segregation to the surface of the cluster is energetically favorable in the Pd–Pt system. Such core–shell segregation is mainly ruled by two factors: maximization of the number of the strongest interatomic interactions and minimization of cluster surface energy.

Experimentally, the surface energy of $Pd \, (125 - 131 \, meV/Å^2)$ is lower than that of $Pt \, (155 - 159 \, meV/Å^2)$ while having a core of Pt atoms also enables the maximization of the number of Pt–Pt interactions (which are the strongest according to both the Gupta potential and DFT). Thus, the DFT calculations show that surface segregation effects of Pd atoms in medium-sized clusters are supported at the higher level of theory, in agreement with previous DFT calculations on Pd–Pt clusters [5].

3.4.3 Analysis of the Dh–cp(DT) Structure

The Dh–cp(DT) structure is best described for the 20–14 composition in which the internal core of Pt atoms is a double tetrahedron or trigonal bipyramid (i.e., two tetrahedra sharing a face). Pd atoms grow on the (111) faces of these two tetrahedra in a regular hcp (111) stacking. On each of the 6 faces of the Pt double tetrahedron (each one formed by 6 Pt atoms), three Pd atoms then grow, giving a total of 18 surface Pd atoms. The two remaining Pd atoms lie on an edge between two faces belonging to the same Pt tetrahedron, thus creating what it is locally a decahedron with its 5-fold axis coinciding with the shared edge. Simultaneously, there is a deformation of the 12 Pd atoms located next to the three edges shared by the two tetrahedra; these Pd atoms minimize their local energy by getting closer to each other and forming three other local decahedral motifs whose axes coincide with the three edges shared by the two tetrahedra. The growth mechanism of the Pd atoms on top of the Pt double tetrahedron is similar to that responsible for the interconversion between 5-fold structural families as theoretically predicted [13, 14] and experimentally verified [15].

Figure 3.7 shows the internal core of the Dh–cp(DT) for all the compositions studied in this work. The peculiar stability of this structure is due to the fact that Pt segregates into the core, where it is highly coordinated and thus preferentially closed-packed, while Pd segregates to the surface, where it is low-coordinated and thus preferentially in decahedral arrangements [7]. The Dh–cp(DT) arrangement thus represents the best compromise for the frustration caused by the different tendencies of the component metals. This is especially true for medium-sized clusters such as those considered in the present work, whereas the crystalline arrangements will eventually prevail for larger clusters [11, 12].

It is interesting to observe that the Dh–cp(DT) structure with the lowest Δ_{34}^{Gupta} and Δ_{34}^{DFT} value is realized at the composition $Pd_{22}Pt_{12}$. In this case, the two extremal (apical) vertices of the Pt double tetrahedron are replaced by two Pd atoms: this lowers the mixing energy of the structure by replacing those internal Pt atoms forming lowest Pt–Pt bonds with Pd atoms (see Fig. 3.7). This reduces strain, as the Pd potential is less "sticky" (i.e. a crystal structure which increases strongly its energy for a change in interatomic distances) than Pt [16]. At the DFT level, there is a further stabilizing contribution due to the fact that the replaced Pt atoms lie in a very asymmetrical bonding environment that is particularly disfavored as Pt has a large orientational (dipolar) energy [17].

Compared to the Dh–cp(DT) arrangement, the TO structure is disfavored because the Pd atoms are not in a decahedral bonding environment, while the "pure" 5-fold non-crystalline structures are disfavored because the internal Pt atoms are not in a closed-packed bonding environment. Finally, the Dh–cp(T) motif is disfavored because (a) there is a smaller proportion of atoms lying on (111) faces in closed-packed stacking and (b) in going from size 32 to size 34, there is a difference between covering the first edge of the double tetrahedron (thus forming decahedral Pd arrangements and releasing strain) or the last edge of the single tetrahedron (thus increasing strain because of competition for decahedra formation with Pd dimers on the other edges).

Fortunelli and co-workers (unpublished work) have performed further calculations on Pd–Pt clusters with a number of atoms in the range 32–44. Preliminary

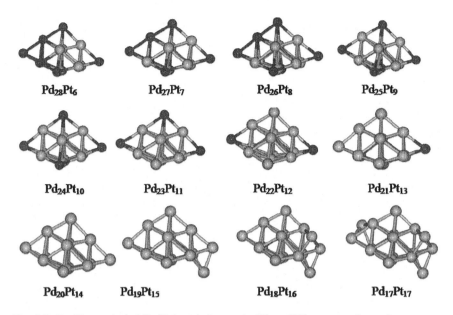

Pd$_{28}$Pt$_6$ Pd$_{27}$Pt$_7$ Pd$_{26}$Pt$_8$ Pd$_{25}$Pt$_9$

Pd$_{24}$Pt$_{10}$ Pd$_{23}$Pt$_{11}$ Pd$_{22}$Pt$_{12}$ Pd$_{21}$Pt$_{13}$

Pd$_{20}$Pt$_{14}$ Pd$_{19}$Pt$_{15}$ Pd$_{18}$Pt$_{16}$ Pd$_{17}$Pt$_{17}$

Fig. 3.7 Double tetrahedral Pt (*light color*) core the Dh–cp(DT) structure for various compositions. At low Pt concentrations, some Pd (*dark color*) atoms are shown for completeness

results indicate that the mixed Dh–cp(DT) motif seems to be preferred in this size range, with a somewhat enhanced stability at size 34 and 36. However, they were unable to single out a cluster in which structural stability was associated with electronic shell closure, as the density of states around the Fermi level always come out rather smooth, confirming the metallic character of these systems.

3.4.4 Conclusions

By combining Gupta potential global optimizations and DFT local energy minimizations [9], we have identified a new particularly stable structural motif, the Dh–cp(DT) structure. In the case of the 34-atom Pd–Pt clusters studied in this work, the energetic ordering of global minimum structural motifs predicted by the Gupta potential is not confirmed at the higher level of theory (i.e., DFT). The Dh–cp(DT) motif is predicted to be the GM only at composition 22–12 at the empirical potential level, whereas it corresponds to the lowest energy structure at the DFT level for all the compositions (17–17 to 28–6) studied here.

This has been rationalized in terms of an optimal compromise between core-segregated, and thus preferentially close-packed, Pt atoms and surface-segregated, and thus preferentially decahedral, Pd atoms. Mixed motifs such as the Dh–cp(DT) structure described in the present work have not been studied much in the literature, whereas they might be quite common for medium-sized binary clusters, which are frustrated because of different bonding tendencies of the two component elements. Finally, the segregation of Pd atoms to the surface of the cluster has been studied at the composition $Pd_{17}Pt_{17}$ and has been corroborated by DFT calculations.

References

1. C. Massen, T.V. Mortimer-Jones, R.L. Johnston, J. Chem. Soc. Dalton Trans. **23**, 4375 (2002)
2. L.D. Lloyd, R.L. Johnston, S. Salhi, N.T. Wilson, J. Mater. Chem. **14**, 1691 (2004)
3. L.D. Lloyd, R.L. Johnston, S. Salhi, J. Comp. Chem. **26**, 1069 (2005)
4. G. Rossi, R. Ferrando, A. Rapallo, A. Fortunelli, B.C. Curley, L.D. Lloyd, R.L. Johnston, J. Phys. Chem. **122**, 194309 (2005)
5. E.M. Fernández, L.C. Balbás, L.A. Pérez, K. Michaelian, I.L. Garzón, Int. J. Mod. Phys. **19**, 2339 (2005)
6. J.P.K. Doye, D.J. Wales, J. Phys. Chem. A **101**, 5111 (1997)
7. R. Ferrando, R.L. Johnston, J. Jellinek, Chem. Rev. **108**, 845 (2008)
8. G. Rossi, A. Rapallo, C. Mottet, A. Fortunelli, F. Baletto, R. Ferrando, Phys. Rev. Lett. **93**, 105503 (2004)
9. G. Barcaro, A. Fortunelli, F. Nita, G. Rossi, R. Ferrando, J. Phys. Chem. B **110**, 23197 (2006)
10. D. Bazin, D. Guillaume, Ch. Pichon, D. Uzio, S. Lopez, Oil Gas Sci. Tech. Rev. IFP **60**, 801 (2005)
11. A.J. Renouprez, J.L. Rousset, A.M. Cadrot, Y. Soldo, L. Stievano, J. Alloy Comp. **328**, 50 (2001)
12. J.L. Rousset, L. Stievano, F.J. Cadete Santos Aires, C. Geantet, A.J. Renouprez, M. Pellarin, J. Catal. **202**, 163 (2001)
13. F. Baletto, R. Ferrando, Rev. Mod. Phys. **77**, 371 (2005)
14. F. Baletto, C. Mottet, R. Ferrando, Phys. Rev. B **63**, 155408 (2001)
15. J.L. Rodríguez-López, J.M. Montejano-Carrizales, U. Pal, J.F. Sánchez-Ramírez, H.E. Troiani, D. García, M. José-Yacamán, Phys. Rev. Lett. **92**, 196102 (2004)
16. F. Baletto, R. Ferrando, A. Fortunelli, F. Montalenti, C. Mottet, J. Chem. Phys. **116**, 3856 (2002)
17. E. Aprà, F. Baletto, R. Ferrando, A. Fortunelli, Phys. Rev. Lett. **93**, 065502 (2004)

Chapter 4
98-Atom Pd–Pt Nanoalloys

4.1 Introduction

In a previous study of Pd–Pt clusters with 1:1 composition [9], it was found that the lowest energy structure for the 56-atom cluster $Pd_{28}Pt_{28}$ has a structure (Fig. 4.1a) which is a fragment of the 98 atom Leary tetrahedron (LT), see Fig. 4.1b, previously discovered by Leary and Doye as the putative global minimum (GM) of the 98-atom Lennard–Jones cluster (LJ_{98}) [5]. The same structure has also been proposed as the lowest energy geometry adopted by 98 atom silver clusters, described with the Sutton–Chen (SC) potential [5], and for a cluster of 98 C_{60} molecules [3]. A subsequent extensive GA search [9] found the full 98 atom LT for the composition $Pd_{49}Pt_{49}$. The objective of the work reported here is to carry out a systematic search of configuration space (using the Gupta potential and combining GA, BHMC and a shell optimisation routines) for 98-atom Pd–Pt clusters, for all compositions Pd_mPt_{98-m}, in order to assess how common the LT structure is for this system. This chapter is fully based on Ref. [11] *Reproduced by permission of the Royal Society of Chemistry.*

4.2 Computational Details

The GA parameters adopted in this study are: population size = 40; crossover rate = 0.8; crossover type = 1-point weighted; selection = roulette; mutation rate = 0.1; mutation type = mutate move; number of generations = 500. Number of GA runs for each composition = 100. As preliminary GA results indicated large oscillations in the plot of excess energy (Δ_{98}^{Gupta}); which were attributed to difficulties in finding low-energy homotops for some compositions, a more detailed homotop search was carried out using the BHMC Monte Carlo algorithm (for a fixed structural

L. O. Paz Borbón, *Computational Studies of Transition Metal Nanoalloys*, Springer Theses, DOI: 10.1007/978-3-642-18012-5_4, © Springer-Verlag Berlin Heidelberg 2011

Fig. 4.1 **a** LT fragment for composition $Pd_{28}Pt_{28}$. **b** LT structure for $Pd_{49}Pt_{49}$. Pd and Pt atoms are represented by *dark* and *light color* spheres, respectively

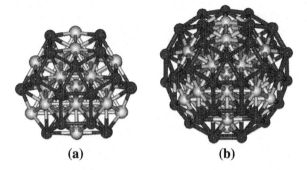

(a)   (b)

configuration and composition), as well as the shell optimization routine for the highly symmetric LT structures (with ideal T_d-symmetry) 98-atom Pd–Pt clusters.

4.3 Results and Discussion

4.3.1 Comparison of Global Minimum Structures for 98-Atom Pd–Pt Nanoalloys

Figure 4.3 shows a plot of Δ_{98}^{Gupta} against Pd content (m) for the putative GM for all compositions Pd_mPt_{98-m} ($m = 0$–98), calculated with the Gupta potential. The lowest values of Δ_{98}^{Gupta} are found in the composition range $Pd_{46}Pt_{52}$ to $Pd_{63}Pt_{35}$, indicating that these are relatively stable GM structures. Over the entire composition range, a number of different structural motifs have been identified as putative GM. These structures, all of which show a certain degree of atomic ordering and core–shell segregation are shown in Fig. 4.2. The GM structures and point group symmetries are listed as a function of composition in Table 4.1. The Marks decahedron (Dh-M: Fig. 4.2b) is predicted by the GA algorithm to be the GM in a wide composition range from Pd_1Pt_{97} to $Pd_{45}Pt_{53}$. Dh Marks structures are denoted by black circles in Fig. 4.3. It should be noted that the structure is actually an incomplete (2, 3, 2) Marks decahedron [11] whose full geometric shell magic number is 101 atoms. Decahedral clusters have previously been reported for 98 atom Ni clusters modelled by the Sutton–Chen (SC) potential [5].

In this composition range, $Pd_{12}Pt_{86}$ is an exception, as it has the same incomplete close-packed tetrahedral structure (cp-T: Fig. 4.2a) as pure Pt_{98}, consisting of a 52 atom truncated tetrahedral Pt core (all four vertex atoms are removed) with the remaining 46 (12Pd + 34Pt, or 46 Pt) atoms forming islands on the (111) faces. Other exceptions are the compositions $Pd_{33}Pt_{65}$, $Pd_{36}Pt_{62}$, $Pd_{39}Pt_{59}$, $Pd_{40}Pt_{58}$, $Pd_{41}Pt_{57}$, and $Pd_{43}Pt_{55}$ where a transition from Dh-Marks to Leary tetrahedron (LT: Fig. 4.2c) structures appears, with no obvious dominance of one motif over the other (LT structures are indicated by white circles in Fig. 4.3).

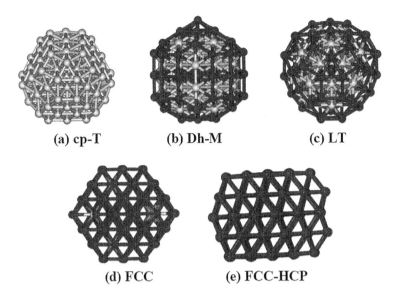

(a) cp-T   (b) Dh-M   (c) LT

(d) FCC   (e) FCC-HCP

Fig. 4.2 Different structural motifs found as GM for 98-atom Pd–Pt clusters

In the composition range $Pd_{46}Pt_{52}$ to $Pd_{63}Pt_{35}$, by using the GA + BHMC atom-exchange approach we found LT as the putative GM structures for all compositions. This range also corresponds to the region of lowest excess energy (Δ_{98}^{Gupta}), see Fig. 4.3 For high Pd concentrations (composition range $Pd_{64}Pt_{34}$ to $Pd_{97}Pt_{1}$), Dh-M structures are again dominant, except for $Pd_{65}Pt_{33}$ (LT) and $Pd_{91}Pt_{7}$ and $Pd_{92}Pt_{6}$ (close packed structures). $Pd_{91}Pt_{7}$ has a mixed fcc-hcp close-packed structure (FCC-HCP: Fig. 4.2e), with ABCBA type stacking of close packed layers, also identified as the GM for Pd_{98}. The putative GM for $Pd_{92}Pt_{6}$ is a fcc close-packed structure (FCC: Fig. 4.2d), which resembles a truncated octahedron. An fcc structure has also been reported for Au_{98} modelled by an SC potential [5].

In Fig. 4.3, all the closed-packed structures (cp-T, FCC and FCC-HCP) are indicated by grey squares. Figure 4.3 also shows the LT isomers with full T_d-symmetry, as obtained by the shell optimisation program. In fact, for many compositions there are several homotops, all with T_d-symmetry, but the figure only shows the lowest-energy T_d-homotop for each composition. The scatter of T_d points is rather oscillatory because the high symmetry homotops are not always the lowest in energy, as they may have an unfavourable placement of Pd and/or Pt atoms. However, for seven compositions ($Pd_{24}Pt_{74}$, $Pd_{36}Pt_{62}$, $Pd_{40}Pt_{58}$, $Pd_{48}Pt_{50}$, $Pd_{52}Pt_{46}$, $Pd_{56}Pt_{42}$ and $Pd_{68}Pt_{30}$), the putative GM is indeed a LT isomer with T_d-shell structure. These T_d-symmetry GM are shown in Fig. 4.4.

For all compositions where LT have been found, the GA finds the LT with much lower probability than alternative metastable Dh-M structures. A careful analysis of GA structures reveals that, even where a LT structure is the putative

Table 4.1 Point group symmetries and structures types of the GM for Pd_mPt_{98-m} clusters

m	Symm	Structure	m	Symm	Structure	m	Symm	Structure
0	C_s	cp-T	33	C_1	LT	66	C_1	Dh-M
1	C_1	Dh-M	34	C_1	Dh-M	67	C_1	Dh-M
2	C_1	Dh-M	35	C_1	Dh-M	68	T_d	LT
3	C_1	Dh-M	36	T_d	LT	69	C_1	Dh-M
4	C_1	Dh-M	37	C_1	Dh-M	70	C_1	Dh-M
5	C_1	Dh-M	38	C_1	Dh-M	71	C_1	Dh-M
6	C_1	Dh-M	39	C_1	LT	72	C_1	Dh-M
7	C_1	Dh-M	40	T_d	LT	73	C_1	Dh-M
8	C_1	Dh-M	41	C_1	LT	74	C_1	Dh-M
9	C_1	Dh-M	42	C_1	Dh-M	75	C_1	Dh-M
10	C_1	Dh-M	43	C_1	LT	76	C_1	Dh-M
11	C_1	Dh-M	44	C_1	Dh-M	77	C_s	Dh-M
12	C_1	cp-T	45	C_1	Dh-M	78	C_1	Dh-M
13	C_1	Dh-M	46	C_1	LT	79	C_1	Dh-M
14	C_1	Dh-M	47	C_1	LT	80	C_1	Dh-M
15	C_1	Dh-M	48	T_d	LT	81	C_1	Dh-M
16	C_1	Dh-M	49	C_1	LT	82	C_1	Dh-M
17	C_1	Dh-M	50	C_{2v}	LT	83	C_1	Dh-M
18	C_1	Dh-M	51	C_{3v}	LT	84	C_1	Dh-M
19	C_1	Dh-M	52	T_d	LT	85	C_1	Dh-M
20	C_1	Dh-M	53	C_{3v}	LT	86	C_1	Dh-M
21	C_1	Dh-M	54	C_{2v}	LT	87	C_1	Dh-M
22	C_1	Dh-M	55	C_{3v}	LT	88	C_1	Dh-M
23	C_1	Dh-M	56	T_d	LT	89	C_1	Dh-M
24	T_d	LT	57	C_1	LT	90	C_1	Dh-M
25	C_1	Dh-M	58	C_1	LT	91	C_1	FCC-HCP
26	C_1	Dh-M	59	C_1	LT	92	C_s	FCC
27	C_1	Dh-M	60	C_1	LT	93	C_1	Dh-M
28	C_1	Dh-M	61	C_1	LT	94	C_s	Dh-M
29	C_1	Dh-M	62	C_1	LT	95	C_1	Dh-M
30	C_1	Dh-M	63	C_1	LT	96	C_1	Dh-M
31	C_1	Dh-M	64	C_1	LT	97	C_s	Dh-M
32	C_1	Dh-M	65	C_1	LT	98	C_1	FCC

GM, only approximately 1% of the GA runs find LT homotops. Dh-M structures are found more frequently, in approx. 20% of GA runs although for some compositions this value decreases. These results are consistent with the results of Leary and Doye for LJ_{98}, where the LT GM was found in only 6 out of 1,000 searches, using a modified BHMC algorithm, with 15 times as many searches ending at the lowest metastable, incomplete icosahedral structure [5].

These results can be attributed to a lack of funnelling towards the LT GM on the potential energy landscape, as previously identified for other non-icosahedral GM structures such as the truncated octahedral GM for LJ_{38} and the decahedral geometry for LJ_{75} [15]. A funnel on a potential or free energy landscape is a region

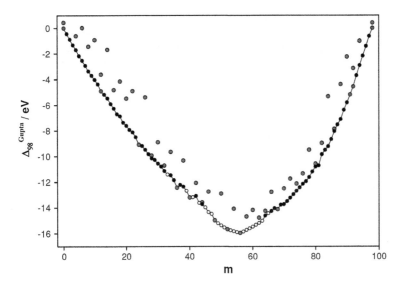

Fig. 4.3 Plot of the excess energy, $\Delta_{98}^{\text{Gupta}}$ as a function of Pd content (m), for 98 atom Pd_mPt_{98-m} clusters, modelled by the Gupta potential. Putative LT GM are denoted by *white circles*, while T_d-symmetry LT, generated by the shell optimization program, are denoted by *red circles*. For each composition, only the lowest energy LT shell structure is shown. The rest of the structural families are denoted by *blue circles* (cp-T, FCC and FCC-HCP) and *black circles* (Dh-M)

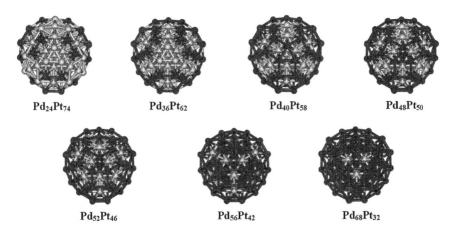

Fig. 4.4 GM LT structures with T_d-symmetry, found using the shell optimization program

of the landscape where there are a number of local minima whose energies decrease systematically towards the centre. The concept of energy funnelling, which has become very important in the study of protein folding [15]. Presumably, on the potential energy landscape of $(Pd, Pt)_{98}$ nanoalloys, there is a greater degree of funnelling towards Dh-M than LT structures. Of course, when relating the success of a search method to the topography of the energy landscape, one has to take into account that the move-class of the search algorithm is also likely to influence the outcome. However, when different search algorithms (e.g. GA and BHMC) find similar ease (or difficulty) in finding the putative GM, for related systems (98-atom Pd–Pt and LJ$_{98}$ clusters), this is more likely to be due to the nature of the landscapes.

4.3.2 Analysis of the Leary Tetrahedron

Leary and Doye have described the Leary tetrahedron structure in terms of a central 20 atom tetrahedron (with fcc packing, Fig. 4.5a), where each of the four (111) faces of the tetrahedron has a truncated (fcc) tetrahedron built on it, forming a 56 atom truncated stellated tetrahedron (Fig. 4.5b). Finally, six puckered centred hexagonal patches are located over the edges of the original tetrahedron, thereby decorating the close-packed surfaces of the stellated tetrahedron and generating the LT, with ideal (if all atoms are equivalent) T_d point group symmetry (Fig. 4.5c) [5]. The 98 atom LT structure in fact is one of a series of Leary tetrahedra [3, 5]: some others are characterised by 34 atoms (e.g. the Gupta potential GM previously found for Pd$_{24}$Pt$_{10}$) [4, 9, 10, 14], 159 and 195 atoms [3, 5]. As noted by Leary and Doye [5] the pseudo-spherical shape of the LT, together with the core, fcc-packing and the high proportion of (111) faces, gives the LT structure a large number (432) of nearest neighbours. They also observed that for LJ$_{98}$ the bulk strain energy of the LT is intermediate between those of the competing icosahedral and decahedral structures. Compared to the FCC and FCC-HCP structural motifs (Fig. 4.2d, e), which are fragments of bulk solids, LT structures provide a better atomic arrangement for these medium-sized clusters.

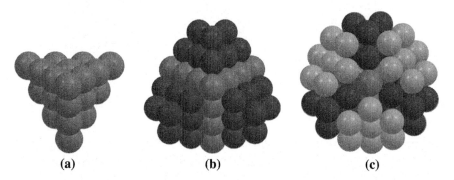

(a)   (b)   (c)

Fig. 4.5 Step by step construction of the LT structure [5] **a** 20 atom tetrahedral core. **b** 56 atom (truncated) stellated tetrahedral unit. **c** Full 98 atom LT

4.3.3 Onion-Like Structures

Analysis of the GM LT structures with the lowest excess energies (as found by the GA + BHMC atom-exchange and shell optimisation approaches) reveals that they possess three-layer Pd–Pt–Pd onion-like geometries, as has previously been observed for icosahedral Pd–Pt clusters [7, 11]. This onion-like geometry may stabilize the LT structure at higher Pd concentrations by increasing the number of (second strongest) Pd–Pt bonds. Figure 4.3 reveals that the lowest $\Delta_{98}^{\text{Gupta}}$ structure, $Pd_{56}Pt_{42}$ (see Fig. 4.4), has full T_d-symmetry (i.e. it is found by the shell optimisation program). A closer inspection of the $Pd_{56}Pt_{42}$ structure reveals that four Pd atoms occupy an inner core tetrahedron (first shell-highly coordinated sites). These 4 atoms are surrounded by 12 Pt atoms (belonging to the second shell of atoms). Another 4 Pd atoms are placed on each of the four vertex positions of the 20 atom tetrahedron, corresponding to surface sites (fifth shell), as expected for Pd atoms, which tend to surface-segregate. 24 Pt atoms form tetrahedral islands on each of the (111) faces of the inner tetrahedron (third and fourth shells). 6 Pt atoms (shell 6) lie over the 6 edges of the 20 atom tetrahedron, constituting the central atoms for the hexagonal Pd rings found on the surface of the cluster. Finally, the remaining 48 Pd atoms cover the surface of the cluster by occupying the low-coordinated sites (shells seven, eight and nine). These findings are consistent with experimental results on Pd–Pt nanoparticles [2, 12, 13] and with the results of previous empirical [6–9, 14] and DFT calculations [4, 10].

Similar building principles are found for the rest of the low $\Delta_{98}^{\text{Gupta}}$ compositions. For example, at composition $Pd_{49}Pt_{49}$, Pd atoms occupy core positions (two Pd atoms) for the first time. Interestingly, for this composition, the structure with a single core Pd atom is found to be higher in energy. In a similar way, the composition $Pd_{50}Pt_{48}$ also has two core Pd atoms. By increasing the concentration of Pd, the number of Pd atoms occupying core positions in the LT structures increases to the maximum of 4 (i.e. the innermost shell of four atoms, defining a tetrahedron). This core is kept for all compositions up to $Pd_{56}Pt_{42}$ (discussed above), which has the lowest value of $\Delta_{98}^{\text{Gupta}}$.

4.3.4 Competition Between Different Structural Families

The energetic competition between the different structural motifs shown in Fig. 4.2 has been investigated in more detail over the composition range $m = 48$–63, which corresponds to the lowest $\Delta_{98}^{\text{Gupta}}$. For compositions in which these motifs were not found by our GA or BHMC searches, the corresponding structures were constructed, and then subjected to BHMC atom-exchange re-optimisation. Figure 4.6 shows a plot of $\Delta_{98}^{\text{Gupta}}$ as a function of composition for these structural motifs. It is interesting to note the close competition between cp-T and Dh-M

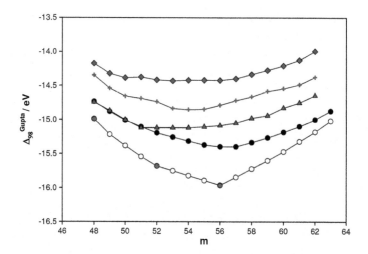

Fig. 4.6 Plot of excess energy Δ_{98}^{Gupta} as a function of Pd content (m) in the range of $m = 48$–63 for various structural motifs: *white circles* (LT); *black circles* (Dh-M); *blue triangles* (cp-T); *blue crosses* (FCC-HCP); and *blue diamonds* (FCC). *Red circles* denote LT shell structures with T_d-symmetry

structures at compositions varying from $Pd_{48}Pt_{50}$ to $Pd_{51}Pt_{47}$, with a difference in total energy between these clusters around 0.01 eV. LT are much lower in energy, by about 0.3 eV, than the cp-T and Dh-M structures. The lowest Δ_{98}^{Gupta} value is reached at composition $Pd_{56}Pt_{42}$ for both LT and Dh-M structures, and at $Pd_{53}Pt_{45}$ for cp-T structures. Figure 4.6 clearly indicates the preference for 98-atom Pd–Pt clusters to adopt LT structures in this composition range. It can also be noted that the excess energies of the closed-packed motifs (cp-T, FCC and FCC-HCP) effectively run in parallel in this composition range, which is reasonable due to their structural similarity. However, the closed-packed structures are not energetically favourable in this composition range.

4.4 Conclusions

Our results indicate that the most stable structures for 98-atom Pd–Pt nanoalloys, modelled by the many-body Gupta potential, adopt the Leary tetrahedron geometry, originally discovered as the global minimum structure for 98 atom Lennard–Jones clusters. LT structures compete with other structural motifs, such as incomplete Marks decahedra and close-packed structures, across the Pd:Pt composition range 1:2–2:1, and are found to dominate in the range $Pd_{48}Pt_{50}$ to $Pd_{63}Pt_{35}$. The high stability of this type of bimetallic (Pd–Pt) nanoalloy can be rationalized in terms of it having an fcc-packed core surrounded by decahedral units, a pseudo-spherical shape, and a large number (432) of nearest neighbours.

Segregation of Pd atoms to surface sites can be explained in terms of the lower surface energy of Pd and the higher bulk cohesive energy of Pt [14]. The combined use of a GA, as an initial global optimisation technique, with the basin hopping atom-exchange approach, and systematic searching for high-symmetry shell structures, has proven to be a powerful technique for studying medium-size bimetallic clusters, especially when there are several competing structural motifs and a very large number of homotops.

References

1. F. Baletto, C. Mottet, R. Ferrando, Phys Rev Lett. **84**, 5544 (2000)
2. D. Bazin, D. Guillaume, Ch. Pichon, D. Uzio, S. Lopez, Oil Gas Sci. Tech. Rev. IFP **60**, 801 (2005)
3. J.P.K. Doye, D.J. Wales, W. Branz, F. Calvo, Phys Rev B **64**, 235409 (2001)
4. E.M. Fernández, L.C. Balbás, L.A. Pérez, K. Michaelian, I.L. Garzón, Int. J. Mod. Phys. **19**, 2339 (2005)
5. R.H. Leary, J.P.K. Doye, Phys Rev E **60**, R6320 (1999)
6. L.D. Lloyd, R.L. Johnston, S. Salhi, J. Comp. Chem. **26**, 1069 (2005)
7. L.D. Lloyd, R.L. Johnston, S. Salhi, N.T. Wilson, J. Mater. Chem. **14**, 1691 (2004)
8. C. Massen, T.V. Mortimer-Jones, R.L. Johnston, J. Chem. Soc. Dalton Trans. 4375 (2002)
9. T.V. Mortimer-Jones, Ph.D. Thesis (University of Birmingham, UK, 2005)
10. L.O. Paz-Borbón, R.L. Johnston, G. Barcaro, A. Fortunelli, J. Phys. Chem. C **111**, 2936 (2007)
11. L.O. Paz-Borbón, T.V. Mortimer-Jones, R.L. Johnston, A. Posada-Amarillas, G. Barcaro, A. Fortunell, Phys. Chem. Phys. **9**, 5202 (2007)
12. A.J. Renouprez, J.L. Rousset, A.M.Cadrot, Y. Soldo, L. Stievano, J. Alloy Comp. **328**, 50 (2001)
13. J.L. Rousset, L. Stievano, F.J. Cadete Santos Aires, C. Geantet, A.J. Renouprez, M. Pellarin, J. Catal. **202**, 163 (2001)
14. G. Rossi, R. Ferrando, A. Rapallo, A. Fortunelli, B.C. Curley, L.D. Lloyd, R.L. Johnston, J. Phys. Chem. **122**, 194309 (2005)
15. D.J. Wales, *Energy Landscapes, with Applications to Clusters, Biomolecules and Glaseses* (Cambridge University Press, Cambridge, 2003)

Chapter 5
38-Atom Binary Clusters

5.1 Introduction

Combined density functional-empirical potential (DFT-EP) approaches are the only viable computational way to sort out the extremely expensive computational calculations for clusters containing more than 20 atoms and have been shown to be particularly efficient when coupled with structural recognition algorithms [1, 2]. The need to explore a wide diversity of structural motifs to make the search of the GM successful has recently been underlined, and approaches such as "system comparison" [3] and hybrid genetic algorithm GA structural + basin-Hopping BHMC atom-exchange [4], among others, which have been proposed to enlarge the data set of structural candidates and to sample the configurational space more thoroughly.

For pure clusters, the competition between crystalline and five-fold symmetric structural motifs, such as decahedra and icosahedra, has been traditionally analyzed [5]. In general, it has been found that non-crystalline arrangements prevail at small sizes, whereas pieces of the bulk crystal progressively become dominant at larger sizes, with the crossover between these motifs being directly related to the "stickiness" of the metal-metal potential [6, 7]. An analysis in terms of structural families is particularly meaningful, as it allows one to roughly predict the order of stability among structural motifs on the basis of a minimal number of calculations. The set of competing structural families has been enlarged to include other possibilities, such as polyicosahedral (pIh) configurations [7, 8], which have been shown to be the GM at both EP and DFT levels for several alloy systems involving size-mismatched second-row and first-row transition elements [9], but which have been predicted to be competitive at the EP level also for second-row/third-row pairs (see, e.g., Refs. [10, 11] and the results depicted in Fig. 5.1 below). A more recent addition is represented by mixed five-fold symmetric/close-packed motifs [4, 12, 13] i.e., motifs exhibiting a close-packed core on which surface atoms grow according to five-fold symmetry axes, that in some cases realize an efficient compromise between the different tendencies of two metallic elements and have

L. O. Paz Borbón, *Computational Studies of Transition Metal Nanoalloys*,
Springer Theses, DOI: 10.1007/978-3-642-18012-5_5,
© Springer-Verlag Berlin Heidelberg 2011

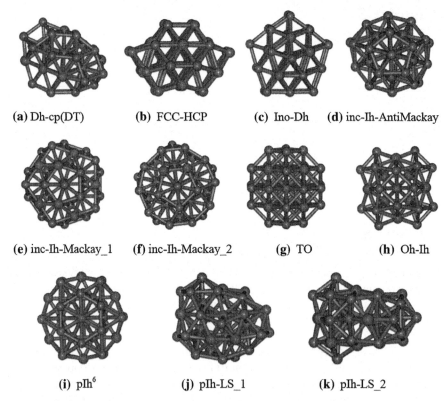

(a) Dh-cp(DT)   **(b)** FCC-HCP   **(c)** Ino-Dh   **(d)** inc-Ih-AntiMackay

(e) inc-Ih-Mackay_1   **(f)** inc-Ih-Mackay_2   **(g)** TO   **(h)** Oh-Ih

(i) pIh^6 **(j)** pIh-LS_1   **(k)** pIh-LS_2

Fig. 5.1 The eleven structural motifs analyzed for 38-atom binary clusters

been predicted to be the lowest-energy structures at the DFT level, as we found for 34-atom Pd–Pt clusters over a wide composition range (see Chap. 3).

We have considered 38-atom binary clusters composed of elements from groups 10 and 11 of the periodic table obtained by combining a second-row with a third-row transition metal (TM), i.e., we consider the four binary TM pairs Pd–Pt, Ag–Au, Pd–Au, and Ag–Pt, and we try to extract useful structural information and insight into the modelling and design of the considered alloy systems from selected calculations combining EP and DFT approaches. Mixed alloys involving noble and quasi-noble metals are interesting both for scientific reasons and in a number of applications, ranging from heterogeneous catalysis to optoelectronic devices [14, 15]. Note that our set includes pairs with a practically zero (Pd–Pt, Ag–Au) or significant (Pd–Au, Ag–Pt) size mismatch, with a modest (Ag–Au, Pd–Au) or large (Ag–Pt, Pd–Pt) difference in the bulk cohesive energies and with differences in electronegativity ranging from large (Ag–Au) to negligible (Pd–Pt) [3]. The pure third-row TMs considered in our study are known to exhibit appreciable directionality effects (see, e.g., Refs. [16–18]) which are difficult to describe using empirical potentials but are here "diluted" by alloying with a second-row TM for which such effects are much less important. Au, especially, but also Pt and Pd, possess a rather "sticky" interaction potential [6, 7], whereas this is less pronounced in the case of Ag. In short, this set allows us to explore in a systematic way a combination of different characteristics. We also observe that from an experimental point of view, mixing second- and third-row transition metals and the corresponding difference in atomic numbers make the investigation of segregation effects for these alloys easier using, for example, X-ray spectroscopy, energy dispersive spectroscopy, and Z-contrast high-angle angular dark-field scanning transmission electron microscopy [15].

5.2 Computational Details

38-atom clusters have been chosen for this study because 38 is a magic number for the truncated octahedral (TO) structure, which has fcc packing (the bulk crystalline structure for all of the metals considered here is fcc) and also for the pIh^6 structure see below, and indeed, the TO or pIh^6 structures are predicted to be the GM for 38-atom clusters by the EP [10, 11]. We will, thus, be able to check whether these structures remain the lowest-energy ones at a higher level of theory or whether mixed five-fold symmetric close-packed or other structures can be competitive at this size and, thus, *a fortiori* in this size range for second-row/third-row transition element pairs in contrast with what is observed for first-row/second-row pairs.

We mainly focused on the specific composition 24–14 (where the first number always refers to the second-row TM atom and the second number to the third-row TM atom) but we also consider selected motifs at compositions 19–19 and 32–6. In addition, the compositions 14–24 and 6–32 are also considered for Pd–Au. The compositions 24–14 and 14–24 are chosen as they are predicted to be at or very close to the minimum in the mixing energy at the EP level see below; i.e., they represent compositions at which the effect of alloying is at its maximum. Compositions 32–6 and 6–32 are included to sample the dependence of our results on this parameter, and 19–19 is of interest for investigating surface segregation effects. Note that 24–14 is also a magic number composition for TO and other structures, while 32–6 is a magic number for TO and pIh^6 structures [10]. To study the energy competition between different structural families and to search for the lowest-energy structural motifs in 38-atom binary clusters, we use our "system comparison" approach [1, 3]. In other words, we first perform systematic GA global optimization runs on the four TM pairs, we collect the lowest-energy structures the putative GM and higher-energy isomers for several compositions around 24–14, and we classify them into structural motifs. We then couple the GA structural search with a BHMC atom-exchange approach [19] in which the putative GM candidates for the four TM pairs are subjected to BHMC-exchange-only calculations, in order to determine the optimal chemical ordering according to the EP. Finally, the lowest-energy configurations belonging to the different structural motifs so derived are subjected to local DFT energy minimization.

In a second step, the problem of the correct chemical ordering at the DFT level is investigated in more detail. This is an important issue as it is known that surface segregation can influence the properties of nanoalloys [14]. The 19–19 composition is selected to investigate this issue as it allows one to compare the energies of configurations obtained by swapping the atomic positions of the two species in the cluster, e.g., from a configuration in which A atoms are in the core and partly on the surface ($A_{core}B_{shell}$) to a configuration in which the B atoms are in the core and partly on the surface ($B_{core}A_{shell}$). Selected structural motifs are considered and local energy optimizations are performed for both the structures with the chemical ordering obtained using the BHMC algorithm and those "inverted homotops" derived from the GM by swapping the positions of the two elements [12]. This allowed one to obtain a quantitative estimate of the driving force to surface segregation at the DFT level.

5.3 Results and Discussion

We initially performed a GA global structural optimization on the four systems for all compositions in order to explore the behaviour of the mixing energy as a function of composition and then we focused mainly on the composition 24–14, chosen because it is exactly at (for the Pd–Pt and Ag–Au pairs) or very close (for Ag–Pt, where the minimum is reached at 25–13) to the minimum of the mixing energy at the EP level, see Fig. 5.2 below. For Pd–Au, the minimum is reached at the composition 14–24, which is, thus, also investigated for this pair, together with the corresponding 6–32 composition. We collected all the lowest-energy structures from the GA searches for all pairs following the "system comparison" philosophy [1, 3]. Eleven structural motifs, shown in Fig. 5.1, were extracted from these searches and chosen as potential GM candidates for DFT relaxation.

The structures were then subjected to BHMC atom-exchange optimization to determine the correct chemical ordering at the EP level for each structural motif [4]. Finally, local geometry relaxations at the DFT level were performed on these structures. Note that imposing the chemical ordering of Pd–Pt for all of the four systems does not qualitatively change the results with respect to using the optimal chemical ordering at the Gupta level for each pair implying differences of about 0.1–0.2 eV. In addition to the composition 24–14, the composition 32–6 was also considered in order to sample compositions with a low concentration of the third-row TM. At this composition, only high-symmetry structures were studied, i.e., TO (O_h), Oh-Ih (D_{4h}), inc-Ih-Mackay-1 (C_{5v}), and a pIh^6 (pancake-type) arrangement with a six-fold symmetry axis (D_{6h}).

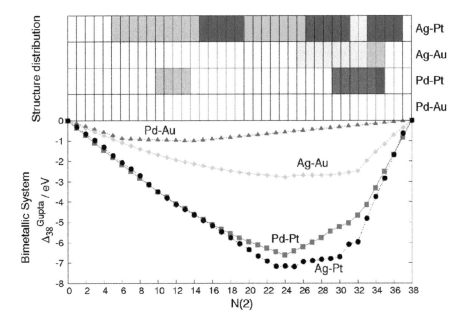

Fig. 5.2 *Bottom* Plot of the Gupta excess energy $\Delta_{38}^{\text{Gupta}}$ for the four bimetallic systems as a function of the number of second-row TM atoms, $N(2)$. *Black circles* are used for Ag–Pt, red squares for Pd–Pt, green diamonds for Ag–Au, and the blue triangles for Pd–Au. *Top* The distribution of GM structures found in our GA search. The colors indicate the different structures: TO (*white*), Ino–Dh (*grey*), inc-Ih-Anti-Mackay (*green*), inc-Ih-Mackay (*magenta*), pIh[6] (*yellow*) and pIh-LS-1 and pIh-LS-2 (*dark* and *light blue*, respectively)

5.3.1 Structural Families

Eleven structural motifs or "families" were singled out via a detailed examination of the GA-EP runs. These structures present a wide range of structural diversity, ranging from fcc crystalline arrangements (TO) to decahedral ones (Ino–Dh), mixed (Oh–Ih, Dh–cp or pIh) and incomplete icosahedral (inc-Ih-Mackay and inc-Ih-AntiMackay) arrangements, among others. These structural motifs are shown in Fig. 5.1 and a detailed description of each is presented here.

(a) Dh–cp(DT): In our previous study of 34 atom Pd–Pt clusters, a novel structural motif was found based on a mixed decahedral/closed-packed arrangement (see Chap. 3) which was identified as the putative GM at the DFT level for 34-atom Pd–Pt clusters, over a wide range of compositions from 17–17 to 28–6. The structure with 38 atoms is simply derived from that found in Chap. 3 by adding two Pd dimers along the edges of the two internal tetrahedral and possesses C_s symmetry. It consists of 14 closed-packed core atoms in a double tetrahedral or trigonal bipyramid arrangement i.e., two tetrahedra sharing a face, further atoms growing on the (111) faces of the two internal tetrahedra in a regular hcp (111) stacking three atoms per face, giving a total of 18 surface atoms, and 6 atoms, four lying on the edges of the same internal tetrahedron and two lying on an edge of the opposite internal tetrahedron. This atomic distribution creates local decahedra with their five-fold axes coinciding with the shared edge of the internal tetrahedra.

(b) and (g) TO and FCC-HCP: TO is the classic fcc-type truncated octahedron O_h symmetry. Its surface presents eight hexagonal faces and six square ones. The fcc–hcp motif is a pseudocrystalline fcc-like structure with a hcp stacking fault and has C_{2v} symmetry. These two structures are similar as they both share an internal octahedron, but the distribution of eight internal atoms in the fcc–hcp structure is somewhat distorted compared to the TO as these atoms are not properly placed on top of each 111 face of the octahedron. See Fig. 5.3 for a graphical description of the core structures of these structures.

(c) Ino–Dh: This is an incomplete Ino-decahedral structure. The complete structure is reached at 39 atoms and has ten rectangular (100) facets. In our case, the lack of one atom just gives nine rectangular facets and C_s symmetry.

(d) inc-Ih-AntiMackay: This structure corresponds to an incomplete anti-Mackay icosahedral structure (the complete structure is reached at size 45). It has an icosahedral (Ih_{13}) core surrounded by 20 atoms lying on top of each (111) face of the internal icosahedron, plus five atoms growing on top of the 12 vertex of the internal icosahedron in a perfect anti-Mackay arrangement, thus, creating five pseudo distorted five-fold axes.

(e) and (f) inc Ih-Mackay-1 and inc Ih-Mackay-2: inc-Ih-Mackay-1 corresponds to an incomplete 55-atom Mackay icosahedron lacking its lower part when viewed from the five-fold axis. It has five square (100) faces and an icosahedral core (Ih_{13}) lacking one atom. inc-Ih-Mackay-2 is similar, with the only difference that the vertex atom lying along the main

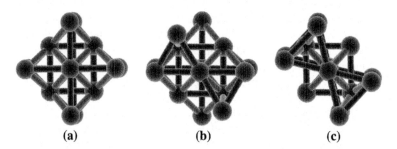

(a)   (b)   (c)

Fig. 5.3 Different atomic arrangements found in similar structures: **a** octahedral core with a perfect growth on top of the (111) and (100) faces, a TO structure. **b** Slight distortion in the fcc growth for the FCC-HCP structure. The distortion leads to octahedral and hexagonal stacking units. **c** The core of the Oh–Ih structure also has an octahedral core, in which the all 8 atoms grow in positions away from top of (111) faces, thus creating what can be classified as two local double-icosahedral arrangements (joined by and octahedral unit)

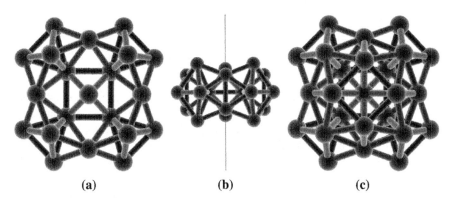

(a)   **(b)**   **(c)**

Fig. 5.4 The Oh–Ih structure (D_{4h}): **a** A view from the top of the internal octahedron, with the 2 square facets (8 atoms) removed. Here it can be appreciated how the two-double icosahedra are joined together. **b** A lateral view of the structure (still with the 8 atoms removed). **c** The complete 38-atom Oh–Ih structure

symmetry axis has been moved to the opposite side of the cluster. Both structures possess C_s symmetry.

(h) Oh–Ih (octahedral–icosahedral): This peculiar structure has not been much considered in the literature before, although it has previously been identified as low-lying (non- GM) isomer for 38-atom homonuclear clusters of Ag, Ni, and Al, using the semi-empirical Gupta (Ag, Ni, see Ref. [20] and Murrell–Mottram (Al see Ref. [21]) potentials. It is similar to the TO structure as it has an internal octahedral core. However, the surface atoms do not grow exactly on top of the (111) faces of the internal octahedron, (as would happen in a TO arrangement) but are placed according to a distorted arrangement, in such a way that only two square faces (instead of six as in the TO) and no hexagonal faces are created. From an internal perspective, it resembles two double icosahedra joined by the central atoms (see Figs. 5.3, 5.4) so that it can be considered as another example of a mixed five-fold-symmetric/closed-packed arrangement, which is being found with increasing frequency in medium-sized binary clusters [4, 12].

(i) pIh^6 (pIh six-fold pancake): 38 atoms represents a "magic" size for completing a six-fold "pancake" structure (D_{6h}) [3, 9, 11, 17]. These structures are found mainly at composition 32–6, where the species that tends to segregate to the surface is found in high concentrations Pd in the case of Pd–Pt, Ag for Ag–Pt and Ag–Au. This structure is known to be favoured among the pIh ones by a small size mismatch between the two metal atoms [3].

(j) and (k) pIh-LS-1 and pIh-LS-2 (pIh low-symmetry structures), These two structures share the same building principle: they both exhibit a quasidouble icosahedron as a pivotal frame with other pseudoicosahedra growing around it. The overall arrangement is rather distorted, hence the name "pIh low-symmetry structures". pIh-LS-2 has a plane of symmetry (C_s), while pIh-LS-1 has no symmetry at all (C_1).

5.3.2 Composition Dependence of the Mixing Energy

Figure 5.2 shows the variation of the Gupta mixing energy $\Delta_{38}^{\text{Gupta}}$ as a function of composition. The negative values of $\Delta_{38}^{\text{Gupta}}$ indicate that the mixing is favoured for all compositions. The minimum in $\Delta_{38}^{\text{Gupta}}$ is reached at compositions 24–14 for Pd–Pt and Ag–Au and at 25–13 for Ag–Pt. For Pd–Au, the minimum occurs at 14–24 i.e., at the Au rich side of the composition range. As will be shown later, this is because Pd–Au is the only system for which the third-row TM atom preferentially segregates to the cluster surface in the EP calculations. Figure 5.2 also shows the GM structural motifs as a function of composition for all four systems (*Structure Distribution*). The TO structure predominates at the EP level, being the GM for all compositions for Pd–Au and for the majority of Pd–Pt and Au–Ag clusters. In these three cases, the minimum in $\Delta_{38}^{\text{Gupta}}$ corresponds to a TO structure. The Ag–Pt system shows the greatest structural diversity, with a preponderance of decahedral, anti-Mackay, and pIh structures, with the minimum in $\Delta_{38}^{\text{Gupta}}$ having an inc-Ih-anti-Mackay structure. In the following, the results of the DFT local optimizations on the EP derived structural motifs are reported and analyzed for each pair.

5.3.3 Pd–Pt Clusters

The relative energies of the 11 structures for the Pd–Pt pair, according to both EP and DFT calculations, are reported in Table 5.1. An analysis of this table shows that the Gupta predictions are in overall fair agreement with the DFT calculations, with the noteworthy exception than at the DFT level, the Oh–Ih motif is lower in energy than the TO structure (predicted to be the GM by the EP), whereas it lies at high energy at the EP level (see Fig. 5.5 Note that the Oh–Ih structure was initially found as a high-energy isomer for the Ag–Au pair, which confirms the usefulness of the system comparison approach [1, 3]. With respect to our previous work at size $N = 34$ (see Chap. 3) we find similar trends, except that now Dh–cp and pIh structures are somewhat destabilized at the DFT level. The Dh–cp structure, in particular, is destabilized by 2 eV relative to the TO. This is due to (a) $N = 38$ being a magic number for TO and Oh–Ih and (b) the addition of two Pd dimers on the edges introduces a strain that was absent for $N = 34$ [12]. pIh structures, such as pIh[6], are destabilized because Pd–Pt presents no size mismatch or electronic shell-closure effects [10, 22]. For this composition and the Oh–Ih structure, Δ_{38}^{DFT} was also evaluated and found to be -4.8 eV, which compares reasonably well with the Gupta value ($\Delta_{38}^{\text{Gupta}} = -6.6$ eV), see Fig. 5.2 and confirms that the Gupta parametrization is reasonably accurate for the Pd–Pt pair.

In Table 5.2, the same quantities are reported at composition 32–6 (higher Pd concentration). The TO structure is now found as the putative GM at the DFT

Table 5.1 Relative energies (in eV) of the 11 structures according to DFT and Gupta calculations for the Pd–Pt, Ag–Pt and Ag–Au pairs, at the 24–14 composition

Structure	Gupta	DFT
	$Pd_{24}Pt_{14}$	
Oh–Ih	0.9282	0.0
TO	0.0	0.1306
Ino–Dh	1.1723	0.9441
fcc–hcp	0.8507	1.0675
inc-Ih-Mackay-1	0.8039	1.1319
Dh–cp(DT)	0.7833	1.3795
pIh-LS-2	1.1622	1.4394
inc-Ih-Mackay-2	1.0810	1.6217
inc-Ih-Anti-Mackay	1.9233	2.4434
pIh^6	1.5826	2.4897
pIh-LS-1	1.4405	2.7563
	$Ag_{24}Au_{14}$	
Ino–Dh	0.5432	0.0
TO	0.0	0.3156
Oh–Ih	0.3552	0.4925
inc-Ih-Mackay-1	0.2084	0.7973
inc-Ih-Anti-Mackay	0.7707	1.1048
inc-Ih-Mackay-2	0.2979	1.1129
fcc–hcp	0.4533	1.6626
pIh-LS-1	0.5511	1.8259
pIh-LS-2	0.4618	2.0653
Dh–cp(DT)	0.4207	2.2966
pIh^6	0.2487	2.9552
	$Ag_{24}Pt_{14}$	
Oh–Ih	1.6474	0.0
inc-Ih-Anti-Mackay	0.0	0.4163
Dh–cp(DT)	0.7595	0.4979
TO	0.8957	0.5524
inc-Ih-Mackay-1	1.0823	0.6367
fcc–hcp	1.4142	0.8054
Ino–Dh	1.2216	0.9116
pIh-LS-1	1.5536	1.2000
inc-Ih-Mackay-2	1.4156	1.2027
pIh-LS-2	1.3946	1.2190
pIh^6	1.7793	1.3932

level, again in close competition (~ 0.04 eV) with the Oh–Ih structure. The other two structures (pIh^6 and inc-Ih-Mackay-1) are destabilized at the DFT level and in the same order as at composition 24–14 but with smaller energy differences, which is reasonable since pure Pd clusters tend to be more fluxional and to exhibit smaller energy differences than pure Pt clusters. Moreover, as Pd and Pt present a

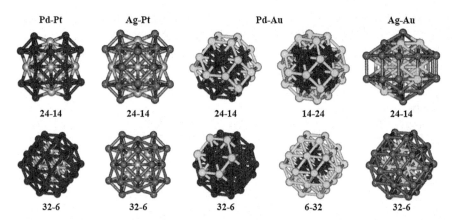

Fig. 5.5 Putative GM structures found at the DFT level for the 24–14 and 32–6 compositions of all four bimetallic systems. For Pd–Au, compositions 14–24 and 6–32 are also shown. Color scheme: Ag and Pd atoms are represented by *dark color* spheres, while Pt and Au atom are indicated by *light color* ones

Table 5.2 Relative energies (in eV) of selected structures according to DFT and Gupta for the Pd–Pt, Ag–Pt, Pd–Au and Ag–Au pairs at the 32–6 composition

	$Pd_{32}Pt_6$		$Ag_{32}Pt_6$		Pd_6Au_{32}		$Pd_{32}Au_6$		$Ag_{32}Au_6$	
	Gupta	DFT	Gupta	DFT	Gupta	DFT	Gupta	DFT	Gupta	DFT
Oh–Ih	0.7473	0.0435	2.2831	0.0	0.2699	0.4598	0.3007	0.1360	1.0132	0.2193
TO	0.2495	0.0	1.8506	0.5768	0.0	0.0	0.0	0.0	0.7078	0.0
inc-Ih-Mackay-1	0.0	0.4272	1.3508	0.1496	0.1753	0.1469	0.2215	0.6530	0.4331	0.4778
pIh^6	0.0986	1.5184	0.0	1.7225	0.9146	1.9130	1.4499	1.5674	0.0	2.2711

small difference in electronegativity (both having a Pauling electronegativity of 2.2), one expects negligible charge transfer between the two metals [3].

5.3.4 Segregation Effects in Pd–Pt Clusters

As Pt has a higher cohesive energy and higher surface energy than Pd, while the atomic radii are practically the same, Pt segregation into the core of the cluster is expected. Indeed, $Pd_{shell}Pt_{core}$ is the preferred chemical ordering according to the EP. This ordering is confirmed by DFT calculations on three different structural motifs at 19–19 composition (see Table 5.3 and Fig. 5.6). Segregation of Pd to surface sites for Pd–Pt clusters has previously been reported on the basis of EP and DFT calculations [12, 23, 24]. It can also be noted that Oh–Ih is definitively at a lower energy than TO at this composition, which implies that Oh–Ih is the putative GM at the DFT level for compositions around 50%–50%.

Table 5.3 Segregation effects at composition 19–19 for the four pairs

	$Pd_{19}Pt_{19}$		$Ag_{19}Pt_{19}$		$Pd_{19}Au_{19}$		$Ag_{19}Au_{19}$	
	EP	Inverted	EP	Inverted	EP	Inverted	EP	Inverted
Oh–Ih	0.0	2.245	0.065	5.861	0.419	5.396	0.446	−0.963
TO	0.729	2.566	0.729	5.758	0.0	5.638	0.0	−0.8630
inc-Ih-Mackay-1	0.811	3.526	0.0	6.221	0.871	6.310	0.078	−0.764

Three structural motifs are investigated at the DFT level (TO, Oh–Ih and inc-Ih-Mackay-1). In each case, two homotops are studied: "EP" is the lowest-energy homotop found using the EP; "Inverted" is the homotop formed by swapping positions of A and B atoms. Both structures are minimized at the DFT level and energies (in eV) are quoted relative to the putative DFT GM. The negative values for $Ag_{19}Au_{19}$ indicate that the EP predicts the wrong segregation for this system

5.3.5 Ag–Pt Clusters

For this pair, as can be seen from Table 5.1, the Gupta potential predicts a strong preference for the inc-Ih-Anti-Mackay structure, with the other isomers well separated in energy. In contrast, at the DFT level, we find that the Oh–Ih motif is the putative GM, as in the case of Pd–Pt system (see Fig. 5.5). It can also be observed that, contrary to what happens for the other TM pairs, DFT calculations shrink instead of magnifying the energy differences, so that one finds several structurally very different motifs at 0.4–0.6 eV above the putative GM. Hence, considerable fluxional behavior is expected for this system at the DFT level. Note that the Dh–cp motif is among the lowest-energy isomers at both the EP and DFT levels: The size mismatch present in this system the atomic radii are ~ 1.445 Å (Ag) and ~ 1.38 Å (Pt) reduces the strain induced by the addition of Ag dimers along the edges of the internal tetrahedra with respect to the Pd–Pt case.

The fact that the Gupta predictions are not confirmed at the higher level of theory is probably due to chemical or directionality effects associated with Pt–Pt interactions [18]. These effects are further enhanced by the size mismatch between the two atoms, which leads to shorter distances and, thus, stronger bonding between the Pt atoms. For this composition and the Oh–Ih structure, Δ_{38}^{DFT} was also evaluated and found to be −2.7 eV. This value is much smaller than the Gupta value ($\Delta_{38}^{Gupta} = -7.1$ eV), see Fig. 5.2 and suggests a reparametrization of the Gupta potential for the Ag–Pt pair. At composition 32–6, one finds similar results (see Table 5.2), with the Oh–Ih motif as the putative GM and inc-Ih-Mackay-1 as a low-energy isomer.

5.3.6 Segregation Effects in Ag–Pt Clusters

In Table 5.3, (see also Fig. 5.7), the total energies for selected structures at composition 19–19 (as predicted by the Gupta potential) and the "inverted" structures obtained by swapping Ag and Pt atoms are reported. In this case, the difference in

Fig. 5.6 Segregation effects
for Pd–Pt clusters at
composition 19–19. Three
structural motifs have been
investigated at the DFT level
(Oh–Ih, TO and inc-Ih-
Mackay-1, from top to
bottom). Energies (in eV) are
quoted relative to the putative
DFT-GM (*top left*). Pd and Pt
atoms are indicated by *dark*
and *light color* spheres,
respectively

EP predicted homotops   **"Inverted" homotop**

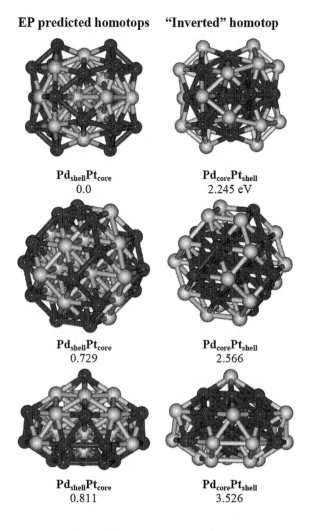

$Pd_{shell}Pt_{core}$ $Pd_{core}Pt_{shell}$
0.0   2.245 eV

$Pd_{shell}Pt_{core}$ $Pd_{core}Pt_{shell}$
0.729   2.566

$Pd_{shell}Pt_{core}$ $Pd_{core}Pt_{shell}$
0.811   3.526

electronegativities between Ag and Pt is not sufficient to favour Pt segregation at the
surface, and the greater strength of Pt–Pt interactions and the larger surface energy of
Pt both combine to favour Ag segregation at the surface, with energy differences for
the inverted homotops of the order of 5–6 eV. It can also be noted that inc-Ih-
Mackay-1 is practically isoenergetic with Oh–Ih and at a lower energy than TO at
this composition (in the same way as at composition 32–6).

5.3.7 Pd–Au Clusters at Compositions 14–24 and 6–32

The Gupta optimizations predict that TO is the preferred structure for all compositions for Pd–Au at size 38, with Au atoms segregating to the surface and Pd

Fig. 5.7 Segregation effects for Ag–Pt clusters at composition 19–19. Three structural motifs have been investigated at the DFT level (Oh–Ih, TO and inc-Ih-Mackay-1, from top to bottom). Energies (in eV) are quoted relative to the putative DFT-GM (*top left*). Ag and Pt atoms are shown in *dark* and *light gray*, respectively

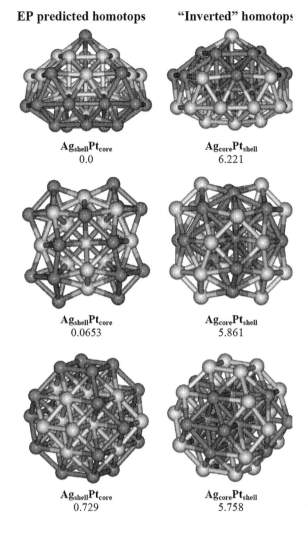

EP predicted homotops   **"Inverted" homotops**

$Ag_{shell}Pt_{core}$
0.0

$Ag_{core}Pt_{shell}$
6.221

$Ag_{shell}Pt_{core}$
0.0653

$Ag_{core}Pt_{shell}$
5.861

$Ag_{shell}Pt_{core}$
0.729

$Ag_{core}Pt_{shell}$
5.758

atoms occupying core positions (see Fig. 5.5). A thorough analysis is shown below for some specific compositions. Due to the interest in Pd–Au nanoparticles in catalytic applications [25, 26], and also due to the very flat behaviour of the mixing energy and the fact that the Gupta potential predicts the lowest mixing energy structures at composition $Pd_{14}Au_{24}$ (see Fig. 5.2), compositions with a high concentration of the third-row atom have also been investigated in this case (see Table 5.4). The EP predicts an unusual surface segregation of the third-row atom for Pd–Au clusters (i.e., the $Pd_{core}Au_{shell}$ arrangement). Our DFT calculations, as shown later for the specific case of the 19–19 composition, are in agreement with the EP prediction. At high Au concentrations, the TO structure is the lowest in energy at DFT level, although the inc-Ih-Mackay-1 structure is also low in energy (ΔE of ~ 0.25 eV), whereas the Ino–Dh, Dh–cp(DT), and pIh structures are destabilized at both the EP and DFT levels. For this composition and the TO structure, Δ_{38}^{DFT} was also evaluated and found to be -2.3 eV, which is somewhat larger than the Gupta value ($\Delta_{38}^{Gupta} = -1.0$ eV), see Fig. 5.2 and points to a fine tuning of the Gupta mixed parameters for the Pd–Au pair. At the 6–32 composition (see Table 5.2), TO is still predicted to be the GM by both approaches, with inc-Ih-Mackay-1 and Oh–Ih as low-lying isomers ($\Delta E \sim 0.1$–0.4 eV).

5.3.8 Pd–Au Clusters at Compositions 24–14 and 32–6

The results of the Gupta potential and DFT calculations for the Pd–Au pair at the 24–14 composition high Pd concentration are reported in Table 5.4. The predicted GM is again a TO structure at both the Gupta potential and the DFT levels, followed by the Oh–Ih and fcc–hcp structures at ~ 0.6 eV and the other motifs at higher DFT energies, in general, amplifies the relative energy differences between the structural motifs, and this is due especially to the fact that pIh structures are strongly disfavored by DFT. This is in line with previous results which indicate that Gupta overestimates the tendency of Au toward five-fold symmetric structures

Table 5.4 Relative energies (in eV) of the 11 structures according to DFT and Gupta calculations for the Pd–Au pair, at the 24–14 and 14–24 compositions

Structure	Gupta	DFT
$Pd_{14}Au_{24}$		
TO	0.0	0.0
inc-Ih-Mackay-1	0.4640	0.2530
inc-Ih-Mackay-2	0.7090	0.4979
fcc–hcp	0.5910	0.8707
Oh–Ih	0.3520	1.2898
Ino–Dh	0.8188	1.6218
Dh–cp(DT)	1.5980	1.9701
inc-Ih-Anti-Mackay	3.0140	1.9864
pIh^6	1.4600	2.0545
pIh-LS-1	1.0700	2.1497
pIh-LS-2	1.2800	2.1796
$Pd_{24}Au_{14}$		
TO	0.0	0.0
Oh–Ih	0.2869	0.6612
fcc–hcp	0.3855	0.6666
Ino–Dh	0.4123	1.4040
inc-Ih-Mackay-1	0.1227	1.5264
inc-Ih-Mackay-2	0.3804	1.6026
Dh–cp(DT)	0.5973	1.6951
pIh-LS-1	0.7428	2.4545
pIh-LS-2	0.9004	2.6041
pIh^6	0.3450	2.6284
inc-Ih-Anti-Mackay	1.4332	2.9250

because of an underestimation of chemical or directionality effects [17, 24]. These expectations are confirmed at composition 32–6 (see Table 5.2), where the TO structure is again predicted to be the putative GM by both approaches but now, icosahedral structures are less disfavored (inc-Ih-Mackay-1 lies at ∼0.6 eV) and Oh–Ih lies at only 0.1–0.2 eV above the putative GM. This is probably due to the smaller size of the Pd atom with respect to the Au atom.

5.3.9   Segregation Effects in Pd–Au Clusters

The DFT total energies for selected structures and their inverted homotops, obtained by swapping Pd and Au atoms, are reported in Table 5.3 (see also Fig. 5.8). The inverted structures are always disfavored, with energy differences of the order of 5–6 eV. This result is easily rationalized. The surface energy of Au is 98 meV/Å^2, whereas that of Pd is 131 meV/Å^2, the cohesive energies of the two species are roughly the same (3.81 vs. 3.89 eV), the atomic radius of Au (1.44Å) is larger than that of Pd (1.38Å), and finally, the Pauling electronegativity of Au (2.4) is slightly larger than that of Pd (2.2) [3], so that any charge transfer from Pd to Au reinforces the tendency to Au surface segregation. It is interesting to note in this respect that experimentally, the prepared Pd–Au particles present gold on the surface, but calcination brings Pd to the surface even after H_2 reduction, probably because of the higher affinity of Pd for oxygen [25]. It can be noted that TO is the putative GM also at this composition, and the energy of inc-Ih-Mackay-1 is intermediate between the values at 14–24 and 24–14 compositions, whereas that of Oh–Ih exhibits more erratic behavior. However, it can be concluded that fcc configurations are definitely favored for Pd–Au particles, with this tendency reinforced around the minimum in the mixing energy, and that, in this case, the EP description compares fairly well with the DFT results.

5.3.10   Ag–Au Clusters

At the Gupta level, the minimum in the mixing energy is reached at composition 24–14 for a perfect TO structure. However, this minimum is not pronounced. Moreover, many structural motifs are predicted by Gupta to lie within an interval of 0.5 eV from the GM (see Table 5.1). As the size mismatch between the two elements is very small (the atomic radii are 1.44Å for Au and 1.445Å for Ag), behaviour similar to Pd–Pt is expected and, indeed, observed at the EP level.

At the DFT level, the major differences are that (a) the energy separation between the different structural motifs is, in general, larger than at the EP level, (b) an Ino–Dh structure becomes the lowest-energy one for $Ag_{24}Au_{14}$ clusters (see Fig. 5.5), and (c) pIh structures are strongly disfavored by DFT. Note that 38 is not

Fig. 5.8 Segregation effects for Pd–Au clusters at composition 19–19. Three structural motifs have been investigated at the DFT level (Oh–Ih, TO and inc-Ih-Mackay-1, from top to bottom). Energies (in eV) are quoted relative to the putative DFT-GM (*top left*). Pd and Au atoms are indicated by *dark* and *light color* spheres, respectively

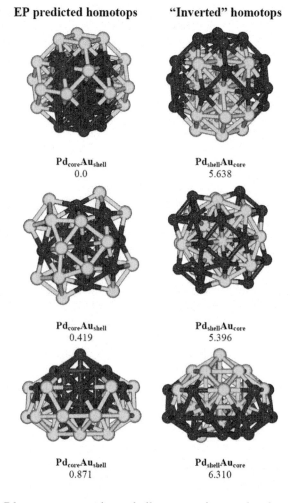

EP predicted homotops   **"Inverted" homotops**

Pd$_{core}$Au$_{shell}$ Pd$_{shell}$Au$_{core}$
0.0   5.638

Pd$_{core}$Au$_{shell}$ Pd$_{shell}$Au$_{core}$
0.419   5.396

Pd$_{core}$Au$_{shell}$ Pd$_{shell}$Au$_{core}$
0.871   6.310

a magic number for the Ino–Dh arrangement whose shell structure is completed at size 39. This reinforces the prediction of this structure as the putative GM in this size range. The TO and the Oh–Ih structures are found as lowest-energy metastable isomers at 0.3–0.4 eV above the Ino–Dh. For this composition and the Ino–Dh structure, Δ_{38}^{DFT} was also evaluated and found to be -1.3 eV, which is smaller than the Gupta value ($\Delta_{38}^{Gupta} = -2.7$ eV), see Fig. 5.2, probably also because of the underestimation of charge transfer effects for the Ag–Au pair.

An analogous explanation can be made for the large energy differences between structural motifs found at the DFT level with respect to the Gupta results, which could also be due to the wrong prediction of segregation effects by the EP. As shown in Table 5.2, at high Ag concentrations (composition 32–6), the TO structure is found as the lowest-energy structure at the DFT level, closely followed by the Oh–Ih structure ($\Delta E \sim 0.20$ eV). However, it is interesting to observe that

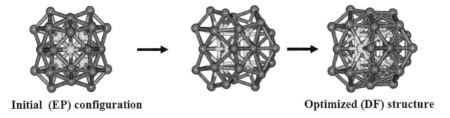

Initial (EP) configuration   Optimized (DF) structure

Fig. 5.9 Schematic representation of the DFT optimization for the Oh–Ih structure of $Ag_{32}Au_6$. The structural distortion, in which Au atoms (*light color spheres*) are displaced toward the surface, can be explained in terms of charge transfer effects

the Oh–Ih structure undergoes a strong $D_{4h} \rightarrow C_s$ distortion after DFT optimization, and some of the Au atoms in the core pop out to the surface (see Fig. 5.9).

5.3.11 Segregation Effects in Ag–Au Clusters

As can be seen from Table 5.3, the most important result obtained in our study of inverted structures is that the $Ag_{shell}Au_{core}$ segregation predicted by the Gupta potential is not confirmed by the DFT calculations. Negative energy differences indicate that the inverted structures $Ag_{core}Au_{shell}$ correspond to the lowest-energy structures at the DFT level (see Table 5.3 and Fig. 5.10). Thus, Au rather than Ag segregation to the surface is predicted by the first-principles approach. The difference must be due to electronic structure effects [17] and, considering that directionality effects associated with the Au atoms are diluted by the presence of Ag atoms and that electronic shell closure is not expected at size 38, the most likely explanation for this unexpected result is charge transfer effects, as underlined in previous works [27–30]. We recall that gold is more electronegative than silver (the Pauling electronegativities are 2.4 and 1.9 for Au and Ag, respectively) [3]. Au atoms can, thus, accept negative charge from silver atoms and, consequently, tend to occupy sites with low coordination, such as surface sites, in order to reduce Coulombic repulsion and charge compression.

To confirm the above interpretation, we have focused on the energies of the semicore $5s$ for Au and $4s$ for Ag levels corresponding to states which are not heavily involved in chemical bonding. These values indicate the direction and the size of the charge transfer because an increase in the electronic density on one atom (negative charge) causes a destabilization of its core levels while increased positive charge leads to core level stabilization. To disentangle electronic and structural effects (which is possible as structural relaxation following Ag–Au exchange is a minor one), we compared the values of the semicore levels for the mixed clusters with those of the pure silver and gold clusters at a frozen geometry [31] (in collaboration with Dr. Giovanni Barcaro and Professor Alessandro Fortunelli). We found that in all the selected 19–19 structures, the Au atoms

EP predicted homotops   **"Inverted" homotops**

Fig. 5.10 Segregation effects for Ag–Au clusters at composition 19–19. Three structural motifs have been investigated at the DFT level (Oh–Ih, TO and inc-Ih-Mackay-1, from top to bottom). Energies (in eV) are quoted relative to the putative DFT-GM (*top left*). Ag and Au atoms are indicated by *dark* and *light color* spheres, respectively

acquire negative charge (i.e., the energies of their $5s$ levels rise), while the silver atoms acquire a corresponding positive charge (i.e., their $4s$ levels drop in energy). Both energy shifts are of the order of 0.5 eV in absolute value and they are opposite in sign.

To perform a deeper analysis, we then focused on a particular structure, the 19–19 TO. TO is a crystalline structure, so size mismatch does not play a role in the stability of a particular chemical ordering (we recall that Au and Ag have very similar atomic radii and bulk lattice constants). Consequently, the two factors that contribute to the optimal chemical ordering are the number of mixed versus pure bonds (i.e., the number of Ag–Au versus Au–Au and Ag–Ag bonds) and charge transfer effects. To separate these two effects and quantify them independently, we have generated several structures characterized by a different chemical ordering and classified them on the basis of the number of pure and mixed bonds. The TO structures have been generated in the following way: The octahedral core is formed by six silver atoms, the atoms at the center of the eight (111) facets are also silver atoms, and the remaining five silver atoms and 19 gold atoms are randomly distributed over the 24 sites constituting the six (100) facets. Structures were selected and DFT local optimizations were performed on them. In all the structures, the total number of bonds is equal to 240 pairs of atoms are considered to form a bond when their separation is less than 3.0Å.

The results of these calculations are reported in Fig. 5.11 where the clusters are classified according to the number of mixed bonds. Note that we have chosen the structures in such a way that if two clusters have the same number of mixed bonds, the numbers of pure Ag–Ag and Au–Au bonds are also the same.

As can be seen from an inspection of Fig. 5.11, the best chemical ordering (lowest energy) is obtained in correspondence with the maximum number of mixed bonds in the cluster (144); the total energy roughly increases when the number of mixed bonds decreases. However, charge transfer effects are also important: An estimate of their size can be made on the basis of the energy spread of clusters characterized by the same number of mixed bonds. From such an analysis, one finds that the maximum variations associated with charge transfer effects are of the order of 0.5–0.7 eV (these values being by chance very similar to the core level shifts of the two species). It, thus, seems necessary for a realistic description of the Ag–Au system to consider additional terms in the EP that take

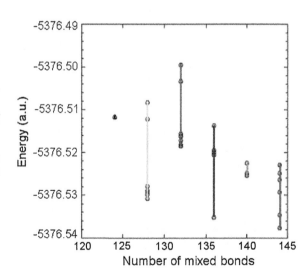

Fig. 5.11 Cluster total energy as a function of the number of mixed bonds for several $Au_{19}Au_{19}$ TO structures. Clusters presenting a variable number of mixed bonds are obtained by permuting the positions of the surface of Ag and Au atoms in the square (100) facets. The energy spread at a fixed number of mixed bonds gives an estimate of the charge transfer effects

into account electrostatic contributions, as recently proposed by Zhang and Fournier [28].

Another factor which can affect the cluster energy is the directionality of bonding. For example, by taking the lowest-energy structure and performing a simple exchange so that a Au atom is moved from a (100) facet to the center of a (111) facet, we found an increase in the energy of 0.33 eV, whereas the exchange between a Au atom at the surface and a Ag atom in the core of the TO produces an increase in energy of only 0.08 eV. This is due to the fact that the directionality contributions in Au strongly destabilize asymmetric environments and are more important for sites of increased coordination, as it happens when a Au atom moves from the corner of a (100) facet to the center of a (111) facet.

5.4 Conclusions

The application of a combined EP-DFT system comparison approach allowed us to single out reasonable candidates for the lowest-energy structures at selected compositions for 38-atom clusters of four bimetallic systems (Pd–Pt, Ag–Pt, Pd–Au, and Ag–Au). From the analysis of the results, the following main conclusions can be drawn:

(1) In general, the EP shows a preference for the TO arrangement except for the Ag–Pt pair, for which an inc-Ih-Anti-Mackay is predicted as the GM. This tendency is confirmed by DFT for Pd–Au (at both 24–14 and 14–24 compositions), but an Ino–Dh arrangement is found as the lowest-energy structure for Ag–Au, while a novel Oh–Ih mixed arrangement is found as a low-energy isomer for Pd–Pt and as the lowest-energy structure for Ag–Pt. The fact that this occurs at $N = 38$, which is a magic number for TO, strongly suggests that mixed five-fold symmetric/close-packed or decahedral arrangements are favored in this size range for the Pd–Pt, Ag–Pt, and Ag–Au pairs.

(2) The EP energy differences between the structural motifs are generally amplified by DFT for the Ag–Au and Pd–Au (at both 24–14 and 14–24 compositions pairs), but the opposite is found for the Ag–Pt pair, while there is no significant trend for the Pd–Pt pair.

(3) pIh structures, which have been proven to be dominant for several pairs involving second-row and first-row transition elements [10] and are predicted to be important competitors by the EP for some of the second-row/third-row transition element binary pairs here investigated, are disfavored by DFT for the Ag–Au, Pd–Pt and Pd–Au pairs at the 24–14 composition (the only exception is the Ag–Pt pair and is associated with a strong fluxional character of the potential energy surface of this system), while the pIh[6] structure is strongly disfavored for all the bimetallic pairs at the 32–6 and 6–32 compositions.

(4) Directionality effects for Au and Pt [17], although diluted by the presence of the second-row metal, subtly modulate the energy ordering, thus, producing differences between EP and DFT results.

(5) The DFT prediction of segregation effects usually coincides with that from the EP calculations. The relative energy differences are found to be of the order of ~ 2 eV for Pd–Pt or 5–6 eV for Ag–Pt and Pd–Au. However, DFT calculations predict Au rather than Ag segregation at the surface in Ag–Au particles because of charge transfer effects (as confirmed by an analysis of the energies of semicore levels for Ag–Au and pure Ag and Au clusters), even though the corresponding energy differences are not large (of the order of 0.5–0.8 eV) and other effects, such as mixing and directionality effects, have been proven to be roughly of the same size.

(6) The difference between Gupta and DFT mixing energies suggests the need for reparametrization of the mixed Gupta interaction parameters especially for the Ag–Pt pair and confirms the importance of including electrostatic contributions for the Ag–Au pair [28].

It is interesting to underline that a peculiar Oh–Ih structure (originally found as a high-energy isomer for Ag–Au and pure Ag clusters [20]) corresponds to the putative GM for two of the systems studied in the present work Pd–Pt and Ag–Pt at composition 24–14. This confirms the usefulness of the system comparison approach in searching for the lowest-energy structures of nanoalloy clusters [1, 3]. Note that the Oh–Ih and TO structures both have an internal octahedral core unit, but surface atoms do not grow in a fcc arrangement in Oh–Ih which is, thus, an example of the mixed five-fold symmetric/close-packed motifs which are being found with increasing frequency in medium-sized binary clusters [4, 12]. Their close energetic competition ($\Delta E \sim 0.5$ eV for all four systems) supports the prediction that fcc-based structures are important competitors for 38-atom size clusters, but mixed arrangements can be the preferred configurations in the neighborhood of this size. This can be rationalized in terms of the crossover between fcc and decahedral structures not yet being complete at size 38 [6, 7], so that a compromise is for the "stickier" third-row transition metals to occupy the core, while the "less sticky" second-row transition metals grow on the surface, forming five-fold symmetric motifs. Finally, when charge transfer effects are not strong, segregation is mainly determined by minimization of surface energy and maximization of the strongest bond interactions, with $Pd_{shell}Pt_{core}$, $Ag_{shell}Pt_{core}$, and $Pd_{core}Au_{shell}$ being the preferred segregation for these three systems.

References

1. R. Ferrando, R.L. Johnston, A. Fortunelli, Phys. Chem. Chem. Phys. **10**, 640 (2008)
2. J. Cheng, R. Fournier, Theo. Chem. Acc. **112**, 7 (2004)
3. A. Rapallo, G. Rossi, R. Ferrando, A. Fortunelli, B.C. Curley, L.D. Lloyd, G.M. Tarbuck, R.L. Johnston, J. Chem. Phys. **122**, 194308 (2005)

4. L.O. Paz-Borbón, T.V. Mortimer-Jones, R.L. Johnston, A. Posada-Amarillas, G. Barcaro, A. Fortunelli, Phys. Chem. Chem. Phys. **9**, 5202 (2007)
5. L.D. Marks, Rep. Prog. Phys. **57**, 603 (1994)
6. F. Baletto, R. Ferrando, A. Fortunelli, F. Montalenti, C. Mottet, J. Chem. Phys. **116**, 3856 (2002)
7. J.P.K. Doye, D.J. Wales, R.S. Berry, J. Chem. Phys. **103**, 4234 (1995)
8. See the Cambridge Cluster Database. http://www-wales.ch.cam.ac.uk/CCD.html
9. G. Rossi, A. Rapallo, C. Mottet, A. Fortunelli, F. Baletto, R. Ferrando, Phys. Rev. Lett. **93**, 105503 (2004)
10. G. Rossi, R. Ferrando, A. Rapallo, A. Fortunelli, B.C. Curley, L.D. Lloyd, R.L. Johnston, J. Phys. Chem. **122**, 194309 (2005)
11. B.C. Curley, R.L. Johnston, G. Rossi, R. Ferrando, Eur. Phys. J. D. **43**, 53 (2007)
12. L.O. Paz-Borbón, R.L. Johnston, G. Barcaro, A. Fortunelli, J. Phys. Chem. C **111**, 2936 (2007)
13. R.H. Leary, J.P.K. Doye, Phys. Rev. E **60**, R6320 (1999)
14. R. Ferrando, R.L. Johnston, J. Jellinek, Chem. Rev. **108**, 845 (2008)
15. R.L. Johnston, R. Ferrando (eds.), Nanoalloys: from theory to applications. Faraday Discuss **138**, 1–441 (2008)
16. E. Aprà, R. Ferrando, A. Fortunelli, Phys. Rev. B **73**, 205414 (2006)
17. R. Ferrando, A. Fortunelli, G. Rossi, Phys. Rev. B **72**, 085449 (2005)
18. E. Aprá, A. Fortunelli, J. Phys. Chem. A **107**, 2934 (2003)
19. J.P.K. Doye, D.J. Wales, J. Phys. Chem. A **101**, 5111 (1997)
20. K. Michaelian, N. Rendón, I.L. Garzón, Phys. Rev. 2000 (1999)
21. L.D. Lloyd, R.L. Johnston, C. Roberts, T.V. Mortimer-Jones, Chem. Phys. Chem. **3**, 408 (2002)
22. G. Barcaro, A. Fortunelli, F. Nita, G. Rossi, R. Ferrando, J. Phys. Chem. B **110**, 23197 (2006)
23. C. Massen, T.V. Mortimer-Jones, R.L. Johnston, J. Chem. Soc. Dalton Trans. **23** 4375 (2002)
24. E.M. Fernández, L.C. Balbás, L.A. Pérez, K. Michaelian, I.L. Garzón, Int. J. Mod. Phys. **19**, 2339 (2005)
25. J.K. Edwards, B.E. Solsona, P. Landon, A.F. Carley, A. Herzing, C.J. Kiely, G.J. Hutchings, J. Catal. **236** 69 (2005)
26. K. Luo, T. Wei, C.W. Yi, S. Axnanda, D.W. Goodman, J. Phys. Chem. B **109** 23517 (2005)
27. R. Mitrić, C. Bärgel, J. Burda, V. Bonačić-Koutecký, P. Fantucci, Eur. Phys. J. D. **24**, 41 (2003)
28. M. Zhang, R. Fournier, J. Mol. Struct. (Theochem) **762**, 49 (2006)
29. V. Bonačić-Koutecký, J. Burda, R. Mitric, M. Ge, G. Zampella, P. Fantucci, J. Chem. Phys. **117**, 3120 (2002)
30. P. Weis, O. Welz, E. Vollner, M.M. Kappes, J. Chem. Phys. **120**, 677 (2004)
31. J. Jellinek, P.H. Acioli, in *The Chemical Physics of Solid Surfaces Atomic clusters: From Gas Phase to Deposited*, vol. 12, ed. by D.P. Woodruff (Elsevier, Amsterdam, 2007)

Chapter 6
Chemical Ordering of 34-Atom Pd–Pt Nanoalloys

6.1 Introduction

Previous computational studies based on the Gupta many-body atomistic potential [1], coupled with the genetic algorithm search method [2], have been used to study the structures and segregation properties of Pd–Pt clusters of various sizes and compositions (see Chaps. 3, 4, 5). Studies have also been performed on onion-like geometric shell Pd–Pt clusters [3, 4, 5]. Using averaged parameters for the mixed (Pd–Pt) interactions, consistent with the experimental mixing behavior of bulk Pd–Pt alloys [6], the experimentally observed tendency of Pd atoms to segregate to the surface of Pd–Pt nanoalloys [7, 8, 9] has been reproduced in these studies and the empirical potential results have been confirmed by ab initio density functional theory (DFT) calculations [10, 11].

In previous calculations on Pd–Pt nanoalloys using the Gupta potential [3, 5, 6, 10, 12, 13, 14], the heteronuclear Pd–Pt parameters were defined as the mean of the homonuclear parameters. This was based on a study by Massen et al. [6], who showed that averaged parameters (parameter set **I**, whose choice was based on the fact that in the bulk Pd–Pt forms solid solutions at all compositions) gave rise to nanoalloy structures with $Pt_{core}Pd_{shell}$ chemical ordering (in qualitative agreement with experimental studies on Pd–Pt nanoparticles [7, 8, 9]. In the small size range studied ($N \leq 56$ atoms), a greater tendency towards decahedral structures was observed for $(Pd-Pt)_{N/2}$ clusters, compared to the pure Pd and Pt clusters [6]. In this original study, it was also shown that making the repulsive pair energy scaling parameter (A) for the Pd–Pt interaction larger (0.35 eV) than for the homonuclear interactions (parameter set **II**) led to layer-like segregation into two sub-clusters, while making the heteronuclear attractive many-body energy scaling parameter (ζ) larger (3.0 eV) than the homonuclear parameters (parameter set **III**) led to clusters with ordered bcc-like (β-brass) structures [6]. Figure 6.1 compares the structures

L. O. Paz Borbón, *Computational Studies of Transition Metal Nanoalloys*,
Springer Theses, DOI: 10.1007/978-3-642-18012-5_6,
© Springer-Verlag Berlin Heidelberg 2011

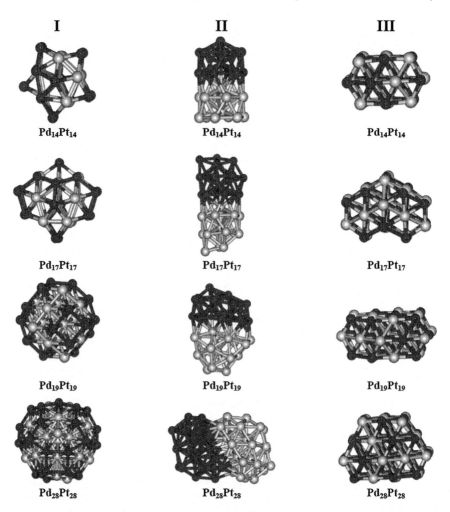

Fig. 6.1 Comparison of structures and chemical ordering of $(Pd-Pt)_{N/2}$ clusters obtained using the Gupta potential parameter sets **I**, **II** and **III**. Pd and Pt atoms are shown as blue and grey spheres, respectively

and segregation patterns for $(Pd-Pt)_{N/2}$ clusters of varying size, using Gupta potential parameter sets **I**, **II** and **III** (see Table 6.1).

Following the initial work of Massen et al. [6], investigating the effect on cluster structure and chemical ordering of making large changes to the energy scaling parameters A and ζ, for fixed 1:1 composition, it was decided to study the effect of more subtle parameter changes. It has been suggested that while it is reasonable that the heteronuclear parameters for bimetallic A–B systems that form solid solutions in the bulk should be intermediate between those of the two pure metals, as found for example by Wang et al. for Co–Cu nanoalloys [15] and by

Table 6.1 Gupta potential parameters used in this study [1, 6]

Parameters	Pd–Pd	Pt–Pt	Pd–Pt(**I-Ia**) $w = 0.5$	Pd–Pt(**II**)	Pd–Pt(**III**)	Pd–Pt(**Ib**) $w = 0.25$	Pd–Pt(**Ic**) $w = 0.75$
A (eV)	0.1746	0.2975	0.23	0.35	0.23	0.27	0.21
ζ (eV)	1.7180	2.695	2.2	2.2	3.0	2.5	2.0
p	10.867	10.612	10.74	10.74	10.74	10.68	10.74
q	3.7420	4.004	3.87	3.87	3.87	3.94	3.87
r_0 (Å)	2.7485	2.7747	2.76	2.76	2.76	2.77	2.76

The w values are the weighting factors used in deriving the Pd–Pt potential parameters, according to Eq. 6.1

Mottet for Ag–Au (C. Mottet, 2008, personal communication), the strength of the A–B interactions should in fact be closer to the stronger homonuclear interactions (say A–A) than to the weaker ones (B–B), as shown by previous studies by Baletto et al. on Ag–Pd clusters (R. Ferrando, G. Rossi, 2007, personal communication) [16]. It was, therefore, decided to test the effect (on structure and chemical ordering) of Pd–Pt parameters derived by taking weighted averages of the Pd–Pd and Pt–Pt parameters, defined as follows:

$$P(\text{Pd–Pt}) = wP(\text{Pd–Pd}) + (1 - w)P(\text{Pt–Pt}) \tag{6.1}$$

where P represents a Gupta potential parameter and w is the weighting factor. Initially, the following weighting factors were investigated: $w = 0.5$ (i.e. the "equal-weighted" averaged parameter set **I** [6], henceforth this set will be labelled **Ia**); $w = 0.25$ ("Pt-biased" parameter set **Ib** corresponding to a higher weighting of the stronger Pt–Pt interactions); and $w = 0.75$ ("Pd-biased" parameter set **Ic**, corresponding to a higher weighting of the weaker Pd–Pd interactions). These parameters are listed in Table 6.1.

A systematic study of the effect of varying the Gupta parameter weighting factor (w) was performed as follows:

(i) 1:1 composition $(\text{Pd–Pt})_{N/2}$ clusters with $N = 2\text{–}20$ atoms were investigated at the Gupta potential level, using parameter sets **Ia**, **Ib** and **Ic**. The putative global minima (GM) structures found with the Gupta potential, using set **Ia** ($w = 0.5$), were subjected to further DFT relaxation. The energetic analysis of Pd–Pt structures (see Fig. 6.2) suggested that their corresponding energies were biased toward Pt–Pt interactions. See Figs. 6.3 and 6.4 for a detailed energy analysis of **Ia** structures used in the DFT optimizations, while Figs. 6.5 and 6.6 correspond for the **Ib** and **Ic** structures, respectively.

(ii) 34-atom $\text{Pd}_m\text{Pt}_{34-m}$ clusters were investigated for all compositions at the Gupta potential level, using parameter sets **Ia**,**Ib** and **Ic**. The results obtained from these calculations (using the Gupta potential) are compared with previously reported Gupta and DFT results (see Chap. 3).

(iii) Finally, the 34-atom $\text{Pd}_m\text{Pt}_{34-m}$ cluster system analysis (at the Gupta potential level) was extended using a finer mesh of weighting parameters (w) ranging from $0 \leq w \leq 1$, in $\Delta w = 0.1$ intervals. Values of the Pd–Pt

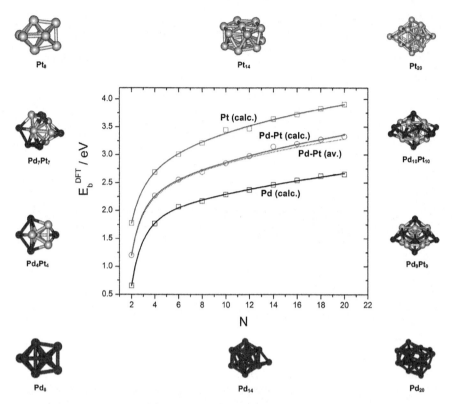

Fig. 6.2 Comparison of DFT binding energies (E_b^{DFT}) for pure Pd, Pt and Pd–Pt clusters with $w = 2$–20. The average of the pure Pd and Pt binding energies, Pd–Pt (av.), are represented by red circles (the red line shows a numerical fit to these values). It is noticeable that the Pd–Pt(calc.) values lie just above the Pd–Pt(av.) curve. Some of the cluster pIh DFT-GM structures (found for pure Pd and Pt, as well as bimetallic Pd–Pt) are shown around the graph

parameters for weighting factors $0 \leq w \leq 1$ are listed in Table 6.1. This chapter is fully based on Ref. [17] (Reproduced by permission of the *Royal Society of Chemistry*).

6.2 Computational Details

6.2.1 Genetic Algorithm

The BCGA GA parameters adopted in this study were: population size = 40 clusters; crossover rate = 0.8 (i.e., 32 offspring are produced per generation); crossover type = 1-point weighted cut-and-splice; selection = roulette wheel; mutation rate = 0.1; mutation type = mutate-move; number of generations = 400. 100 GA runs were performed for each composition.

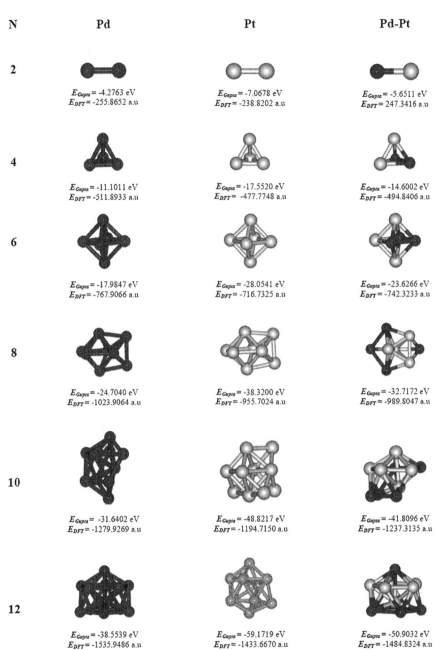

Fig. 6.3 GM structures found by DFT optimizations for Pd_N, Pt_N and $(Pd-Pt)_{N/2}$ for clusters for $N = 2$–12 atoms. Parameter set **Ia** was used for the combined GA-Gupta global optimisation calculations. Gupta potential and Density Functional Theory (DFT) cluster total energies are denoted by E_{Gupta} and E_{DFT}, respectively

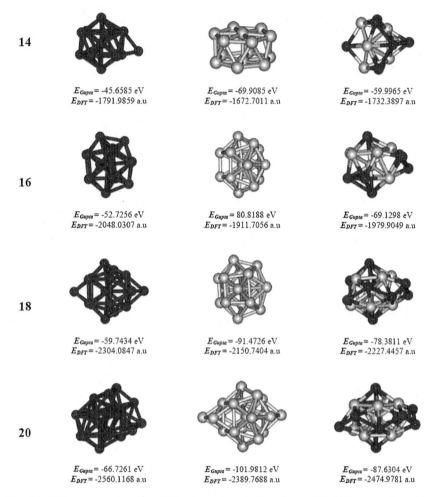

14

E_{Gupta} = -45.6585 eV
E_{DFT} = -1791.9859 a.u

E_{Gupta} = -69.9085 eV
E_{DFT} = -1672.7011 a.u

E_{Gupta} = -59.9965 eV
E_{DFT} = -1732.3897 a.u

16

E_{Gupta} = -52.7256 eV
E_{DFT} = -2048.0307 a.u

E_{Gupta} = 80.8188 eV
E_{DFT} = -1911.7056 a.u

E_{Gupta} = -69.1298 eV
E_{DFT} = -1979.9049 a.u

18

E_{Gupta} = -59.7434 eV
E_{DFT} = -2304.0847 a.u

E_{Gupta} = -91.4726 eV
E_{DFT} = -2150.7404 a.u

E_{Gupta} = -78.3811 eV
E_{DFT} = -2227.4457 a.u

20

E_{Gupta} = -66.7261 eV
E_{DFT} = -2560.1168 a.u

E_{Gupta} = -101.9812 eV
E_{DFT} = -2389.7688 a.u

E_{Gupta} = -87.6304 eV
E_{DFT} = -2474.9781 a.u

Fig. 6.4 GM structures found by DFT optimizations for Pd_N, Pt_N and $(Pd-Pt)_{N/2}$ for clusters for $N = 14$–20 atoms. Parameter set **Ia** was used for the combined GA-Gupta global optimisation calculations. Gupta potential and Density Functional Theory (DFT) cluster total energies are denoted by E_{Gupta} and E_{DFT}, respectively

6.3 Results and Discussion

6.3.1 DFT Study of $(Pd-Pt)_{N/2}$ Clusters $(N = 2$–20)

Figure 6.2 shows the binding energies (E_b^{DFT}) of $(Pd-Pt)_{N/2}$ clusters with $N = 2$–20 atoms, calculated at the DFT level, as well as the corresponding binding energies for pure Pd_N and Pt_N clusters using parameter set $Ia(N = 0.5)$. All the structures of the Pd, Pt and Pd–Pt clusters, with 2–20 atoms, as well as their DFT

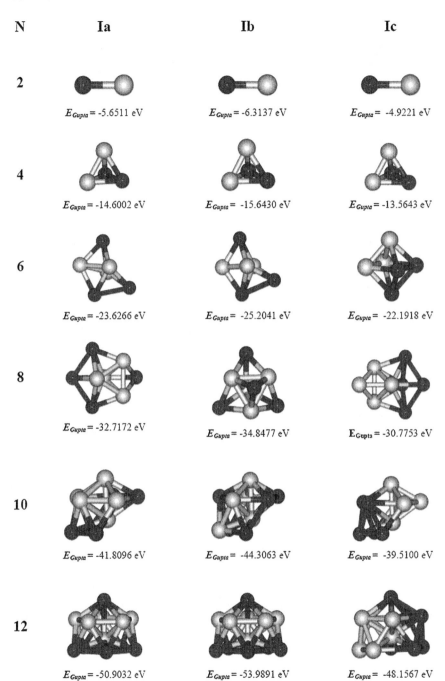

Fig. 6.5 Comparison of structures and chemical ordering in $(Pd-Pt)_{N/2}$ for clusters (for $N = 2-12$) using Gupta parameter sets **Ia** ($w = 0.5$), **Ib** ($w = 0.25$) and **Ic**($w = 0.75$). Gupta potential cluster total energies are denoted by E_{Gupta}

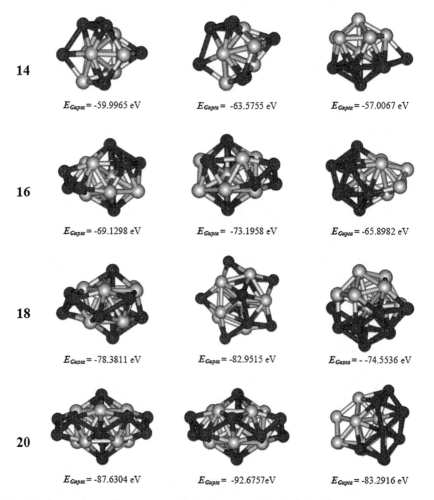

14

$E_{Gupta} = $ -59.9965 eV $E_{Gupta} = $ -63.5755 eV $E_{Gupta} = $ -57.0067 eV

16

$E_{Gupta} = $ -69.1298 eV $E_{Gupta} = $ -73.1958 eV $E_{Gupta} = $ -65.8982 eV

18

$E_{Gupta} = $ -78.3811 eV $E_{Gupta} = $ -82.9515 eV $E_{Gupta} = $ - -74.5536 eV

20

$E_{Gupta} = $ -87.6304 eV $E_{Gupta} = $ -92.6757eV $E_{Gupta} = $ -83.2916 eV

Fig. 6.6 Comparison of structures and chemical ordering in $(Pd-Pt)_{N/2}$ for clusters (for $N = 14$–12) using Gupta parameter sets **Ia** ($w = 0.5$), **Ib** ($w = 0.25$) and **Ic** ($w = 0.75$). Gupta potential cluster total energies are denoted by E_{Gupta}

(E_{DFT}) and Gupta potential (E_{Gupta}) total energies are shown in Figs. 6.3 and 6.4. From the graph in Fig. 6.2 it is apparent that the binding energies for bimetallic $(Pd-Pt)_{N/2}$ clusters (red line) lie slightly above the average of the binding energies of the Pd_N and Pt_N clusters (dashed blue line), i.e. closer to the results for Pt than Pd clusters. This is consistent with the idea proposed above that the bonding in bimetallic clusters should be closer to that of the more cohesive (stronger binding energy) metal component. These lines in Fig. 6.2 have been derived by fitting the binding energies (E_b) to the following equation:

$$E_b = a + bN^{-1/3} + cN^{-2/3} + dN^{-1} \tag{6.2}$$

which recognizes that the binding energy for a cluster of size (N) can be written as the sum of volume, surface, edge and vertex contributions [18].

6.3.2   Gupta Potential Study of $(Pd-Pt)_{N/2}$ Clusters (N = 2–20) for Potentials Ia, Ib and Ic

Figures 6.5 and 6.6 compare the structures and chemical ordering of the GM for several even-numbered bimetallic $(Pd-Pt)_{N/2}$ clusters with N = 2–20 atoms, for Gupta parameter sets **Ia**($w = 0.5$), **Ib**($w = 0.25$), and **Ic**($w = 0.75$). It is clear from Figs. 6.5 and 6.6 that the results for the equal-weighted ($w = 0.5$) and Pt-biased ($w = 0.25$) potentials are quite similar—both in terms of the structures and the preference for $Pd_{shell}Pt_{core}$ segregation. On the other hand, the Pd-biased ($w = 0.75$) potential is more likely to favour different structures (e.g. for N = 6, 14, 18 and 20) and, in all cases, gives rise to very different chemical ordering, corresponding to layer-like (rather than core–shell) segregation, where one half of the cluster is Pd and the other is Pt. It should be noted, however that the segregation observed for potential **Ic** is less severe than in the case previously observed for potential **II** by Massen et al. [6], where the clusters adopted prolate ellipsoidal geometries, with the Pd and Pt atoms effectively segregated into discrete subclusters (see Fig. 6.1).

6.3.3   Gupta Potential Study of 34-Atom $Pd_{34}Pt_{34-m}$ Clusters for Potentials Ia, Ib and Ic

Figure 6.7 compares the excess energies (Δ_{34}^{Gupta}), calculated using Gupta potentials **Ia**, **Ib** and **Ic**, for the putative GM of 34-atom (Pd_mPt_{34-m}) clusters, as a function of the number of Pd atoms (m). Table 6.2 shows the excess energy values for each set of parameters. Several features of Fig. 6.7 are noteworthy. Firstly, the curve for the Pd-biased potential (**Ic**) has positive Δ_{34}^{Gupta} values for all compositions, corresponding to endothermic Pd–Pt mixing (i.e. enthalpy favours segregation of the Pd and Pt atoms), while the curves for the equally weighted (**Ia**) and Pt-biased (**Ib**) potentials have negative Δ_{34}^{Gupta} values for all compositions, corresponding to exothermic Pd–Pt mixing.

Secondly, the exothermicity of mixing (more negative Δ_{34}^{Gupta} values) is greatest for the Pt-biased potential, since Pt–Pt interactions are stronger than Pd–Pd interactions. Finally, the shapes of the **Ia** and **Ib** curves are different, with the **Ib** (Pt-biased) curve having a clear single minimum at (Pd,Pt) composition (22,12), while the **Ia** (equal-weighted) curve has a minimum at composition (21,13), though it is fairly flat in the composition range (21,13)–(24,10). The shape of the

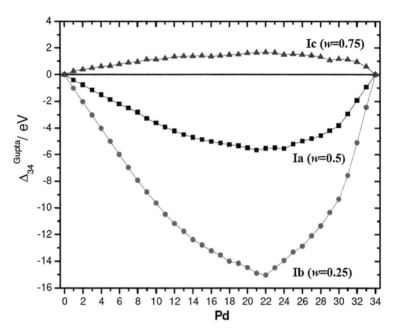

Fig. 6.7 Excess energy plot as a function of Pd concentration (m) for 34-atom Pd–Pt clusters, using Gupta potential parameter sets **Ia**, **Ib** and **Ic**

excess energy curve, at the Gupta level, for the equal-weighted averaged potential has been noted previously [10] and corresponds to a region where the GM structure is very sensitive to the cluster composition. The structures of the putative GM of 34-atom $Pd_{34}Pt_{34-m}$ clusters are compared in Fig. 6.8, for potentials **Ia**, **Ib** and **Ic**, for selected compositions. As was observed for the smaller clusters, again the equal-weighted (**Ia**) and Pt-biased (**Ib**) potentials favour $Pd_{shell}Pt_{core}$ segregation, while the Pd-biased (**Ic**) potential favours layered segregation. These segregation patterns are consistent with the excess energy curves plotted in Fig. 6.7, with the endothermic curve for the **Ic** potential corresponding to de-mixing.

The structures identified as the GM for the equal-weighted averaged potential (**Ia**) agree with those previously reported [10], with the exception of composition (26,9), for which, after performing a considerably greater number of GA runs, we found a Dh–cp(DT) motif to be lower in energy than the previously reported Dh–cp(T) motif [10]. Our results show a significant composition dependence. The structures stabilised by the Pt-biased potential also show considerable dependence on composition. While there is sometimes a degree of similarity between the GM arising from potentials **Ia** and **Ib**, it is noticeable that the structures in the vicinity of the minimum (19,15) to (25,9) in Δ_{34}^{Gupta} for **Ib** are all based on the capped anti-Mackay icosahedron (also known as the 5-fold polyicosahedron), as found for compositions (20,14) and (21,13) for potential **Ia** [10], though for compositions (19,15) and (20,14) the anti-Mackay arrangement is somewhat distorted. This continuity of structure probably explains the smoother behaviour of the Δ_{34}^{Gupta} curve around the minimum for the **Ib**

Table 6.2 Excess energy $\Delta_{34}^{\text{Gupta}}$ values for weighted **Ia**, **Ib** and **Ic** Gupta potential parameters for GM Pd_mPt_{34-m} clusters

m	$w = 0.5$ (**I = Ia**)	$w = 0.25$ (**Ib**)	$w = 0.75$ (**Ic**)
0	0.00000000	0.000000000	0.000000000
1	−0.39641671	−1.019655706	0.264288294
2	−0.76775941	−2.026945412	0.378817588
3	−1.15111712	−3.030833118	0.490562882
4	−1.51217182	−4.022190824	0.610860176
5	−1.86420553	−5.009603529	0.661611471
6	−2.20773024	−5.990828235	0.779132765
7	−2.50359994	−6.971715941	0.895492059
8	−2.81744565	−7.927060647	0.954013353
9	−3.28038235	−8.797667353	1.131546647
10	−3.61876306	−9.643890059	1.138858941
11	−3.94094976	−10.48694776	1.214300235
12	−4.23176347	−11.17044247	1.353801529
13	−4.49413418	−11.76210118	1.382632824
14	−4.71513488	−12.37177488	1.391854118
15	−4.88588459	−12.77864259	1.414842412
16	−5.02678129	−13.21403429	1.372577706
17	−5.13845600	−13.53047800	1.429707000
18	−5.25458171	−13.99967871	1.484369294
19	−5.34055741	−14.15174641	1.526441588
20	−5.49545412	−14.46693212	1.629235882
21	−5.66665282	−14.89698082	1.644608176
22	−5.53617353	−15.02800753	1.681854471
23	−5.49079624	−14.49249824	1.643643765
24	−5.55269794	−13.93692694	1.503246059
25	−5.22516865	−13.29677665	1.552889353
26	−4.98501235	−12.86299735	1.495101647
27	−4.80734106	−12.09242106	1.442535941
28	−4.58618976	−11.34401676	1.341705235
29	−4.21342647	−10.35142447	1.078722529
30	−3.83459218	−9.344242176	1.184145824
31	−2.94599188	−7.560528882	1.132399118
32	−1.91988359	−5.111056588	0.948076412
33	−0.94285929	−2.446978294	0.605170706
34	0.00000000	0.000000000	0.000000000

potential (see Fig. 6.8). It is also worth noting that, although both potentials **Ia** and **Ib** predict high symmetry (C_{5v}) anti-Mackay homotops for $Pd_{21}Pt_{13}$, the Pt-biased potential (**Ib**), with its greater preference for Pd–Pt mixing, has a Pd-centred Pd_1Pt_{12} icosahedral core, surrounded by an anti-Mackay dodecahedral Pd_{20} shell, with one of the pentagonal faces of the dodecahedron capped by a Pt atom, giving a Pd:Pt:Pd:Pt onion-like structure. On the other hand, the equal-weighted potential (**Ia**) has a greater tendency towards core–shell segregation, yielding a lowest energy homotop

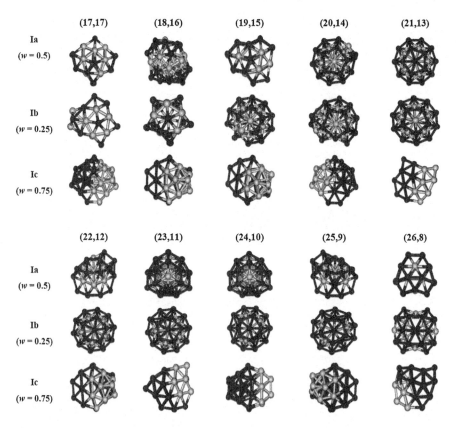

Fig. 6.8 Comparison of the structures of the putative GM of 34-atom Pd_mPt_{34-m} clusters for potentials **Ia**, **Ib** and **Ic** for selected compositions

consisting of a Pt_{13} icosahedron, surrounded by a Pd_{20} anti-Mackay shell, which is in turn capped by a Pd atom.

6.3.4 Detailed Study of 34-Atom Pd_mPt_{34-m} Clusters for Variable Gupta Potential Weights

In view of the results presented before it was decided to carry out a more detailed investigation of the structures and chemical ordering for all compositions of 34-atom Pd–Pt (Pd_mPt_{34-m}) clusters, with the parameter weighting factor w ranging from 0 [i.e. $P(Pd–Pt) = P(Pt–Pt)$] to 1 [i.e. $P(Pd–Pt) = P(Pd–Pd)$] in steps $\Delta w = 0.1$ (see Table 6.3 for a detailed description). 100 GA runs were carried out for each (m, w) point.

Figure 6.9 (top) shows a map of the structural motifs of the putative GMs as a function of composition (m) and weighting factor (w). In agreement with Fig. 6.7, Fig. 6.9 (bottom) shows that the excess energy Δ_{34}^{Gupta} is most negative (most

Table 6.3 Weighted Pd–Pt Gupta potential parameters

Parameters	$w = 0.0$	$w = 0.1$	$w = 0.2$	$w = 0.3$	$w = 0.4$	$w = 0.5$
A (eV)	0.2975	0.28521	0.27292	0.26063	0.24834	0.23605
ζ (eV)	2.695	2.5973	2.4996	2.4019	2.3042	2.2065
p	10.612	10.6375	10.663	10.6885	10.714	10.7395
q	4.004	3.9778	3.9516	3.9254	3.8992	3.873
r_0 (Å)	2.7747	2.77208	2.76946	2.76684	2.76422	2.7616

	$w = 0.6$	$w = 0.7$	$w = 0.8$	$w = 0.9$	$w = 0.4$	$w = 1.0$
A (eV)	0.22376	0.21147	0.19918	0.18689	0.1746	
ζ (eV)	2.1088	2.0111	1.9134	1.8157	1.718	
p	10.765	10.7905	10.816	10.8415	10.867	
q	3.8468	3.8206	3.7944	3.7682	3.742	
r_0 (Å)	2.75898	2.75636	2.75374	2.75112	2.7485	

exothermic mixing) for the highest Pt bias ($w = 0$) and is most positive (most endothermic mixing) for the highest Pd bias ($w = 1$). The lowest excess energy structures for $0 \leq w \leq 0.6$ (exothermic mixing) and the highest excess energy structures for $0.7 \leq w \leq 1$ (endothermic mixing) are shown in Fig. 6.10. See Tables 6.4 and 6.5 for a full numerical description of the calculated mixing energy ($\Delta_{34}^{\text{Gupta}}$) for each w, as a function of composition, Pd_mPt_{34-m}.

The highlighted area (black rectangle) shown in the top part of Fig. 6.9 shows the intermediate composition and weighting factor (m, w) range of main interest in this study, as the optimal (lowest excess energy) compositions lie in the range $Pd_{17}Pt_{17}$ to $Pd_{26}Pt_8$ (for $0.4 \leq w \leq 0.8$) and weighting factors outside this range give too much Pd or Pt bias. This region is interesting as all of the five different structural motifs are found here. We expanded the GM search by carrying out 300 GA runs for each (m, w) point in this region.

6.3.5 Structural Motifs

After performing the global optimisations using the Gupta potential for all compositions of 34-atom Pd–Pt clusters, as well as for all w factors, we were able to categorize the GM structural motifs into five different structural families, with cell colourings (as shown in Fig. 6.9, top) as follows: poly-icosahedra (pIh, blue); incomplete Marks decahedra (Dh, white); decahedra with close-packed double tetrahedral cores (Dh–cp(DT), green); decahedra with close-packed tetrahedral cores (Dh–cp(T), yellow); and incomplete truncated octahedra (TO, red).

Poly-icosahedra (pIh). For Pt-biased w parameters in the range $0 \leq w \leq 0.4$, the majority of the GM structures share the same building principle: they adopt structures based on interpenetrating icosahedra. As shown in Fig. 6.10, the lowest excess energy structures in this w range have 13-atom centred icosahedral cores, which are surrounded (in an anti-Mackay fashion) by 20 triangular-face-capping

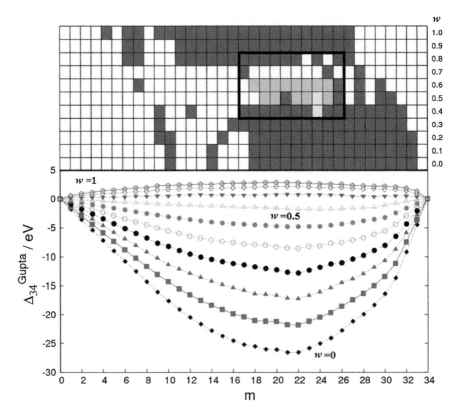

Fig. 6.9 (*Top*) Map of the GM structural motifs found for 34-atom Pd–Pt clusters, as a function of composition (number of atoms of Pd, *m*) and weighting factor (*w*). Different colours indicate particular structural motifs: Dh (*white*), pIh (*blue*), Dh–cp(DT) (*green*), incomplete TO (*red*) and Dh–cp(T) (*yellow*). The preference for GM Dh–cp(DT) motifs at *w* = 0.6, in agreement with our work on 34-atom DFT results is noticeable (see Chap. 3). The *black rectangle* highlights the area of interest for the study of segregation, as shown in Fig. 6.11. (*Bottom*) Comparison of excess energies (Δ_{34}^{Gupta}) for 34-atom Pd$_m$Pt$_{34-m}$ clusters for different values of weighting-parameter (*w*)

atoms, with one of the resulting pentagonal faces being capped to generate the final 34-atom cluster, the highest possible symmetry of which is C_{5v}. For *w* factors in the Pd-biased regime ($0.8 \leq w \leq 1.0$) the poly-icosahedral structures are found to have layered-type segregation, having a distorted arrangement with no symmetry.

Decahedra (Dh). Incomplete Marks decahedral units are found distributed over a wide composition range for Pd–Pt clusters. They are also found to be the GM structures for pure Pd and Pt clusters. It is noticeable that decahedral motifs tend to stabilize Pd–Pt clusters at low Pd concentrations over a wide *w* range ($0 \leq w \leq 0.7$), where mixing is favoured. Decahedral structures are also found for high Pd concentrations in the region ($0.7 \leq w \leq 1.0$) where layered segregation dominates.

Mixed decahedral-close-packed motifs Dh–cp(DT) and Dh–cp(T). The Dh–cp(DT) structure is best described at the (20,14) composition [10] for which the internal core of 14 Pt atoms is a double tetrahedron (or trigonal bipyramid),

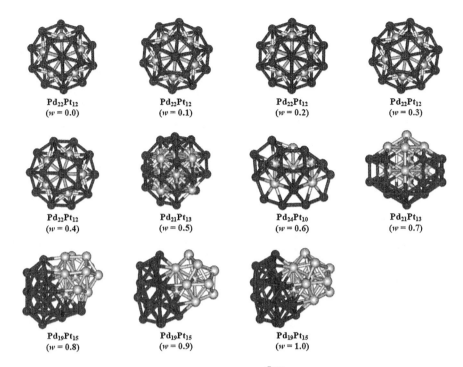

Pd$_{22}$Pt$_{12}$
($w = 0.0$)

Pd$_{22}$Pt$_{12}$
($w = 0.1$)

Pd$_{22}$Pt$_{12}$
($w = 0.2$)

Pd$_{22}$Pt$_{12}$
($w = 0.3$)

Pd$_{22}$Pt$_{12}$
($w = 0.4$)

Pd$_{21}$Pt$_{13}$
($w = 0.5$)

Pd$_{24}$Pt$_{10}$
($w = 0.6$)

Pd$_{21}$Pt$_{13}$
($w = 0.7$)

Pd$_{19}$Pt$_{15}$
($w = 0.8$)

Pd$_{19}$Pt$_{15}$
($w = 0.9$)

Pd$_{19}$Pt$_{15}$
($w = 1.0$)

Fig. 6.10 Comparison of the lowest excess energy (Δ_{34}^{Gupta}), found for weighting factors in the range of $0 \leq w \leq 0.6$ and maximum excess energy structures for $w \geq 0.6$

i.e., two tetrahedra sharing a face. Core segregation of Pt is favoured by the higher cohesive energy and lower surface energy of Pt. Pd atoms grow on the (111) faces of these two tetrahedra in a regular hcp (111) stacking. On each of the six faces of the Pt double tetrahedron (each one formed by six Pt atoms), three Pd atoms can grow, giving a total of 18 surface Pd atoms. The two remaining Pd atoms lie on an edge between two faces belonging to the same Pt tetrahedron, thus creating what it is locally a decahedron, with its 5-fold axis coinciding with the shared edge. On the other hand, the Dh–cp(T) motif corresponds to a highly symmetrical structure (T_d symmetry at composition (24,10)), which has a perfect tetrahedral core of 10 Pt atoms surrounded by the remaining 24 Pd atoms.

Incomplete truncated octahedra (TO). These structures are fragments of fcc packing, based around an octahedral core. For the specific case of Pd$_{28}$Pt$_6$, at $w = 0.7$, the internal core is formed by six Pt atoms, with the Pd atoms occupying 28 of the possible 32 sites of a perfect truncated octahedron.

In our previous study of 34-atom Pd–Pt clusters (see Chap. 3), for all compositions, using the averaged Pd–Pt Gupta parameter set **(Ia)**, [10] we found that in the (Pd,Pt) composition range (17,17)–(28,6) the lowest energy structural motif changed rapidly with composition: (incomplete Marks) Dh [(17,17) and (19,15)]; (low symmetry) pIh [(18,16)]; (anti-Mackay) pIh [(20,14) and (21,13)];

Table 6.4 Excess energy $\Delta_{34}^{\text{Gupta}}$ values for weighted Gupta potential parameters in the range $0 \leq w \leq 0.5$ for GM Pd_mPt_{34-m} clusters

m	$w = 0.0$	$w = 0.1$	$w = 0.2$	$w = 0.3$	$w = 0.4$	$w = 0.5$
0	0.00000	0.00000	0.00000	0.00000	0.00000	0.00000
1	−1.79142	−1.47095	−1.16642	−0.87657	−0.60028	−0.33651
2	−3.58284	−2.93717	−2.32308	−1.73807	−1.17985	−0.65607
3	−5.37425	−4.40238	−3.47733	−2.59502	−1.75300	−0.97144
4	−7.16565	−5.86088	−4.62047	−3.43860	−2.32130	−1.28434
5	−8.95609	−7.31733	−5.76038	−4.27747	−2.88414	−1.58136
6	−10.74479	−8.76966	−6.89511	−5.11229	−3.43714	−1.87457
7	−12.52764	−10.22125	−8.02942	−5.94000	−3.96017	−2.21678
8	−14.30708	−11.65340	−9.14032	−6.74330	−4.45701	−2.50066
9	−16.06414	−13.05941	−10.23037	−7.63559	−4.99134	−2.78404
10	−17.60753	−14.29236	−11.36188	−8.19134	−5.48853	−3.05553
11	−19.10822	−15.52921	−12.13633	−8.91368	−5.97764	−3.32681
12	−20.50207	−16.57235	−12.88340	−9.52917	−6.44978	−3.56974
13	−21.67322	−17.52306	−13.58245	−10.03839	−6.78903	−3.79644
14	−22.79106	−18.38291	−14.24218	−10.43920	−7.09939	−4.00066
15	−23.74606	−19.11625	−14.82350	−10.83268	−7.34474	−4.15183
16	−24.54731	−19.86792	−15.37552	−11.17346	−7.52510	−4.27221
17	−25.23610	−20.41653	−15.78646	−11.42202	−7.78994	−4.37586
18	−25.53841	−21.04733	−16.39914	−11.75333	−7.89050	−4.47331
19	−26.03168	−21.18634	−16.49859	−12.04466	−8.00555	−4.56907
20	−26.09738	−21.23602	−16.68455	−12.29376	−8.29140	−4.69758
21	−26.55824	−21.78451	−17.15663	−12.70022	−8.44009	−4.79261
22	−26.57585	−21.84368	−17.26219	−12.83837	−8.58048	−4.78921
23	−25.83920	−21.18414	−16.68397	−12.34688	−8.18274	−4.75719
24	−25.04467	−20.48530	−16.08051	−11.83905	−8.11093	−4.74070
25	−24.11176	−19.67140	−15.38312	−11.25753	−7.65555	−4.54639
26	−22.75698	−18.54469	−14.48636	−11.04478	−7.54132	−4.34291
27	−21.20212	−17.43587	−13.83396	−10.39491	−7.12754	−4.10255
28	−19.96086	−16.33121	−12.96750	−9.76349	−6.72599	−3.78906
29	−18.33189	−15.00139	−11.82823	−8.91413	−6.15234	−3.53801
30	−16.40381	−13.42515	−10.67173	−8.05266	−5.57114	−3.23240
31	−13.68559	−11.14764	−8.71933	−6.48679	−4.41256	−2.43763
32	−9.18050	−7.47271	−5.88056	−4.36109	−2.91420	−1.60890
33	−4.47310	−3.63488	−2.83389	−2.08323	−1.42440	−0.79599
34	0.00000	0.00000	0.00000	0.00000	0.00000	0.00000

Dh–cp(DT) [(22,12)]; Dh–cp(T) [(23,11)–(25,9)]; TO [(26,8)]; and (incomplete 6-fold pancake) pIh [(27,7) and (28,6)]. Interestingly, at the DFT level the situation was found to be less complicated, with the Dh–cp(DT) structure being preferred over the entire composition range (17,17)–(28,6) [10]. The DFT calculations also confirmed the Gupta prediction of $Pd_{shell}Pt_{core}$ chemical ordering.

Considering the structural motifs identified in the present study (Fig. 6.9, top), and looking at the line for $w = 0.5$ (corresponding to the averaged parameters), we can see that for Pt-rich compositions Dh structures are lowest in energy, while at

Table 6.5 Excess energy $\Delta_{34}^{\text{Gupta}}$ values for weighted Gupta potential parameters in the range $0.6 \le w \le 1$, for GM Pd_mPt_{34-m} clusters

m	$w = 0.6$	$w = 0.7$	$w = 0.8$	$w = 0.9$	$w = 1.0$
0	0.00000	0.00000	0.00000	0.00000	0.00000
1	−0.08434	0.15398	0.37325	0.58344	0.78515
2	−0.22161	0.18600	0.56660	0.91968	1.24456
3	−0.28128	0.25316	0.71948	1.14317	1.52476
4	−0.36887	0.29817	0.91358	1.45430	1.88362
5	−0.48226	0.29953	1.00902	1.56315	1.99187
6	−0.59855	0.34125	1.14354	1.74824	2.27903
7	−0.71856	0.48990	1.26841	1.94913	2.46203
8	−0.76579	0.46549	1.31547	1.97461	2.55667
9	−0.88615	0.54264	1.55288	2.28341	2.74216
10	−0.96344	0.50796	1.64604	2.32902	2.80515
11	−1.03358	0.64497	1.74857	2.42602	2.86710
12	−1.10589	0.69870	1.85228	2.44645	2.87390
13	−1.21829	0.75185	1.90479	2.59586	3.11954
14	−1.24560	0.69202	1.82439	2.52110	3.12557
15	−1.30612	0.70055	1.89678	2.67120	3.23024
16	−1.38298	0.67897	1.91477	2.74178	3.21560
17	−1.48845	0.66891	1.98327	2.75584	3.27023
18	−1.48984	0.66115	2.05465	2.77958	3.28404
19	−1.55045	0.71312	2.16094	2.85924	3.41633
20	−1.62211	0.75605	2.14123	2.84390	3.34081
21	−1.74369	0.81561	2.14652	2.83701	3.36468
22	−1.81764	0.75379	2.12162	2.77726	3.30457
23	−1.82622	0.73806	2.05820	2.68990	3.21505
24	−1.83312	0.73342	1.94882	2.58491	3.13431
25	−1.76966	0.72260	1.93740	2.57952	3.08968
26	−1.73288	0.67697	1.97049	2.53331	3.02244
27	−1.70248	0.48103	1.82387	2.35329	2.79489
28	−1.55336	0.48669	1.68982	2.27892	2.79689
29	−1.49234	0.39379	1.77025	2.19165	2.56341
30	−1.12938	0.45664	1.61427	2.18298	2.54501
31	−0.78882	0.51992	1.62154	2.06762	2.43501
32	−0.45927	0.50220	1.38160	1.87084	2.23700
33	−0.19782	0.34956	0.85651	1.33698	1.79142
34	0.00000	0.00000	0.00000	0.00000	0.00000

the Pd_{rich} extreme (27,7)–(34,0) Dh and pIh structures compete. However in the region highlighted in the black rectangle [(17,17)–(26,8)] the Dh–cp(DT) structural motif is predominant, with exceptions being: (17,17) and (18,16) [Dh]; (21,13) and (26,8) [TO]; and (24,10) [Dh–cp(T)].

It should be noted that in this study the Gupta potential parameters for each value of w have been rounded to four decimal places. Because of this, our results for $w = 0.5$ differ slightly from those previously reported using the averaged parameter set **Ia** [10]. This shows the sensitivity of the potential to slight changes in the numerical values of the Gupta potential parameters around $w = 0.5$. Interestingly, the new $w = 0.5$ parameter set appears to be in better agreement with the previous DFT calculations.

For weighting factor $w = 0.6$ (i.e. slightly Pd-biased Pd–Pt parameters), the Dh–cp(DT) motif is found as the Gupta potential GM for a wider composition range (18,16)–(25,9), which is in better agreement with the previous DFT results [10]. The excess energy curves in Fig. 6.9 (bottom) confirm that Pd–Pt mixing is still energetically favoured for $w = 0.6$ (i.e. excess energies are all negative) consistent with the observed core–shell chemical ordering (see Fig. 6.10). Of course, biasing the Pd–Pt interactions slightly towards the weaker (Pd–Pd), rather than the stronger (Pt–Pt), interactions goes against previous considerations [18] and perhaps indicates a competition between structural and energetic factors. Interestingly, moving to Pt-biased potentials ($w < 0.5$) leads to a predominance of pIh structures in this composition range, while going to more Pd-biased potential ($w > 0.6$) stabilizes first Dh and then pIh structures. The Dh–cp(DT) structures are, therefore, only found at the Gupta potential level for $w = 0.5$ and 0.6.

6.3.6 Segregation

The lowest excess energy structures for each value of w, shown in Fig. 6.10, illustrate that for low w values (Pt-biased potentials) core–shell ordering is observed, while for high w values (Pd-biased potentials) layer-like spherical cap segregation is favoured. (It should be noted that this spherical cap segregation has been previously discussed by Christensen et al. in a study of the size dependence of phase separation in small Cu–Ag and other bimetallic clusters [19].) For $w = 0.7$, an intermediate "ball-and-cup" segregation is found. These types of segregation are shown schematically in Fig. 6.11 along with a map of segregation types in the region of Pd compositions ($m = 17$–26) and weighting factors ($w = 0.4$–0.8) highlighted by the black rectangle in Fig. 6.9 (top). In agreement with Fig. 6.10, core–shell segregation was found for average and Pt-biased potentials ($w \leq 0.6$) and spherical cap segregation for Pd-biased potentials ($w \leq 0.8$), with intermediate ball-and-cup segregation for intermediate ($w \sim 0.7$), as shown schematically in Fig. 6.11. Segregation was also found to depend on composition, with core–shell configurations being stabilized at higher w, for higher Pd concentrations. The segregation patterns are consistent with the excess energy plots shown in Fig. 6.9 (bottom), where endothermic Pd–Pt mixing (positive excess energy) is found for all compositions for weighting factors $w = 0.7$.

6.3.7 Study of Interfacial Areas in Segregated Bimetallic Clusters

In the previous section it was shown that with increasing weighting factor w (i.e. biasing the Pd–Pt Gupta potential parameters towards the weaker Pd–Pd

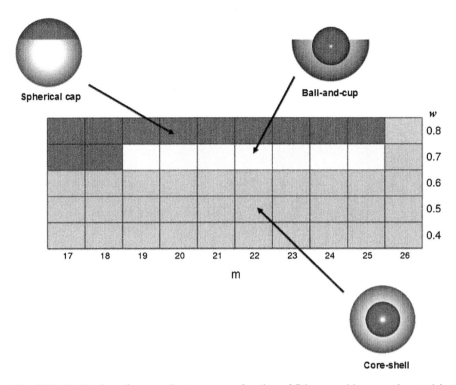

Fig. 6.11 Distribution of segregation types as a function of Pd composition m and potential weighting factor w; core–shell (*blue*); ball-and-cup (*cream*); spherical cap (*red*), respectively

interactions) Pd–Pt mixing becomes unfavoured and spherical cap segregation, rather than core–shell ordering, is observed. In this section, we show in a general way that this occurs so as to minimize the interfacial area between the two subsets of atoms (say A and B) composing the bimetallic cluster.

Figure 6.12 shows a plot of the A–B interfacial areas for the limiting cases of core–shell and spherical cap (layered) segregation, for bimetallic clusters of composition $A_x B_{1-x}$, as a function of the mole fraction ($x = x_A$) of the core- or cap-forming species A. In the following analysis, we have assumed that the overall cluster shape is spherical (with a cluster radius of R) and that the atomic densities of the two metals are the same ($\rho_A = \rho_B$), so the volumes of the two parts of the segregated clusters ($V_A = V_B$) can be obtained as: $V_A = xV_{sph}$ and $V_B = (1 - x)V_{sph}$, where the volume of the cluster sphere $V_{sph} = \frac{4}{3}\pi R^3$.

For core–shell segregation, the interfacial area is the surface area of the spherical core, $A_{core} = 4\pi r_{core}^2$, where r_{core} is the radius of the core. For spherical-cap segregation, the interfacial area is the area of the circular base of the cap, $A_{cap} = 4\pi r_{cap}^2$, where r_{cap} is the radius of this circle. For the 50:50 composition ($x = 0.5$), the interfacial area for the core–shell configuration is $A_{core} = 2^{4/3}\pi R^2$, while for the spherical cap configuration, $x = 0.5$ corresponds to the special case

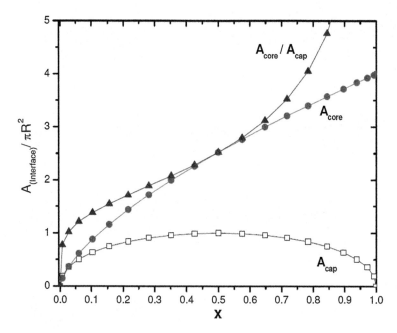

Fig. 6.12 Variation of interfacial area for the core–shell (A_{core}) and spherical cap (A_{cap}) segregated clusters and the ratio A_{core}/A_{cap} as a function of the mole fraction x of the core/cap element

where the cluster is divided into two equal hemispheres, so the interfacial area A_{cap} is the cross sectional area of the sphere (πR^2).

Figure 6.12 shows that the plot of A_{cap} vs. x is symmetrical about $x = 0.5$ (with the species A merely changing from the minority to the majority species). However, A_{core} increases with x, such that in the limit of $x = 1$, the core has expanded to be the whole cluster (i.e. $A_{core} = 4\pi R^2$). Figure 6.12 also shows the ratio $\eta = A_{core}/A_{cap}$, which tends to 0 as $x \rightarrow 0$ (as A_{core} falls more rapidly than A_{cap} at very low A concentrations) and goes to ∞ as $x \rightarrow 1$ (as $A_{cap} = 0$ at $x = 1$, whilst $A_{core} = 4\pi R^2$). The plot of η shows that, apart from for very small A compositions ($x < 0.025$), $A_{core} > A_{cap}$, and that at 0.5, $\eta = 2^{4/3} \sim 2.52$ (Table 6.6).

The observed tendency (see Figs. 6.10 and 6.11 for spherical cap layered segregation for high (i.e. Pd-biased) Pd–Pt weighting factors (w), therefore, occurs in order to lower the Pd–Pt interfacial area. This de-mixing is also consistent with the calculated positive excess energy (see Fig. 6.9), indicating unfavourable Pd–Pt mixing. It should be noted that although the core–shell configuration is regarded as segregated, the relatively large interfacial area of the core–shell clusters means that there are a high number of Pd–Pt interactions, which is why such structures are favoured by w values ranging from around 0.6 (averaged) towards 0 (Pt-biased), for all of which the excess energy is negative indicating favourable Pd–Pt mixing. The ball-and-cup configuration is an intermediate between the

Table 6.6 Numerical values for x, A_{cap}, A_{core} and the ratio A_{core}/A_{cap}, for $\Delta\gamma = 0.1$

x	$\dfrac{A_{cap}}{\pi R^2} = \gamma(2 - \gamma)$	$\dfrac{A_{core}}{\pi R^2} = 4x^{2/3}$	$\dfrac{A_{core}}{A_{cap}} = \dfrac{4x^{2/3}}{\gamma(2 - \gamma)} = \eta$
0	0	0	0
0.007	0.19	0.148	0.779
0.028	0.36	0.368	1.022
0.061	0.51	0.620	1.216
0.104	0.64	0.884	1.381
0.156	0.75	1.160	1.547
0.216	0.84	1.440	1.714
0.282	0.91	1.720	1.890
0.352	0.96	1.996	2.079
0.425	0.99	2.260	2.283
0.500	1.00	2.520	2.520
0.575	0.99	2.764	2.792
0.648	0.96	2.996	3.121
0.718	0.91	3.208	3.525
0.784	0.84	3.400	4.048
0.844	0.75	3.572	4.763
0.896	0.64	3.716	5.806
0.939	0.51	3.836	7.522
0.972	0.36	3.924	10.900
0.993	0.19	3.980	20.947
1	0	4	∞

extremes of the more mixed core–shell and the more segregated spherical cap configurations. Finally, it should be noted that the layered segregation previously found by Massen et al. [6] using Pd–Pt parameter set **II** (see the structures reproduced in Fig. 6.1) is similar to the spherical cap segregation observed here. However, the segregation in those earlier studies was more extreme, with the clusters adopting prolate ellipsoidal, rather than pseudo-spherical, geometries in order to further reduce the interfacial area between the Pd and Pt atoms.

6.3.8 Derivation of the Interfacial Area Formulae

6.3.8.1 Core–Shell Segregation

In Fig. 6.13, assuming $\rho_{core} = \rho_{shell}$, $R = R_{sph}$ and $r = r_{core}$

$$x = x_{core} = \frac{V_{core}}{V_{sph}} = \frac{\frac{4}{3}\pi r^3}{\frac{4}{3}\pi R^3} = \frac{r^3}{R^3} \tag{6.3}$$

and $A_{core} = 4\pi r^2 = 4\pi R^2 \cdot x^{2/3}$.

For $x = 0.5$, we have $r = \frac{R}{2^{1/3}}$ and $A_{core} = \frac{4\pi R^2}{2^{2/3}} = 2^{4/3}\pi R^2$.

Fig. 6.13 Definition of
geometrical parameters for
core–shell segregated clusters

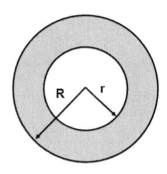

6.3.8.2 Spherical Cap Segregation

Considering Fig. 6.14, we have $r^2 = 2hR - h^2$, and $A_{cap} = \pi r^2 = \pi(2hR - h^2)$; where $R = R_{sph}$, $r = r_{cap}$ and $h = $ height of cap, then:

$$V_{cap} = \frac{\pi}{6}(3r^2 + h^2)h = \frac{\pi}{3}h^2(3R - h) \tag{6.4}$$

Defining $h = \gamma R$, with $0 \leq \gamma \leq 1$ corresponding to the composition range $0 \leq x \leq 0.5$, A_{cap} and V_{cap} can be expressed in terms of R and γ:

$$A_{cap} = \pi R^2 \gamma (2 - \gamma) \tag{6.5}$$

$$V_{cap} = \frac{\pi}{3}R^3 \gamma^2 (3 - \gamma) \tag{6.6}$$

and

$$x = x_{cap} = \frac{V_{cap}}{V_{sph}} = \frac{\frac{\pi}{3}R^3 \gamma^2 (3 - \gamma)}{\frac{4\pi}{3}R^3} = \frac{\gamma^2 (3 - \gamma)}{4} \tag{6.7}$$

For a given value of x, γ is obtained by solving the cubic equation $\gamma^3 - 3\gamma^2 + 4x = 0$ numerically. Although this equation can be shown to have real roots, only one of them lies in the physically reasonable range $0 \leq \gamma \leq 2$. For the special case of $x = 0.5$ (where the spherical cap is a hemisphere), we have:

Fig. 6.14 Definition of
geometrical parameters for
spherical-cap segregated
clusters

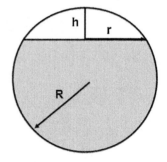

$$r_{\text{cap}} = R \qquad (6.8)$$

$$A_{\text{cap}} = \pi R^2 \qquad (6.9)$$

In the composition range $0.5 < x \leq 1$ (i.e. $1 < \gamma \leq 2$ and $R < h \leq 2R$), the minority and majority species are interchanged but the plot of $A_{\text{cap}}(x)$ is symmetrical about $x = 0.5$; i.e. $A_{\text{cap}}(\gamma) = A_{\text{cap}}(2 - \gamma)$, with $x(\gamma) = 1 - x(2 - \gamma)$.

6.3.8.3 Ratio of Interfacial Areas

In general, the ratio between interfacial areas can be expressed as:

$$\eta = \frac{A_{\text{core}}}{A_{\text{cap}}} = \frac{4\pi R^2 x^{2/3}}{\pi R^2 \gamma(2 - \gamma)} = \frac{4x^{2/3}}{\gamma(2 - \gamma)} \qquad (6.10)$$

For the special case of $x = 0.5$ ($\gamma = 1$):

$$\eta = \frac{a\pi R^2 \left(\frac{1}{2^{2/3}}\right)}{\pi R^2} = 2^{4/3} \approx 2.52 \qquad (6.11)$$

6.4 Conclusions

We have presented a detailed study of the structures and chemical ordering of Pd–Pt nanoalloy clusters as a function of composition and the strength of the heteroatomic interactions, which has been achieved by defining the Pd–Pt parameters of the Gupta many-body potential energy function as a weighted average of the weaker (Pd–Pd) and stronger (Pt–Pt) homoatomic parameters, with the weighting factor (w) ranging from 0 (Pt-biased) to 1 (Pd-biased).

An initial DFT energetic analysis of small Pd–Pt clusters ($N = 2$–20), with 50:50 compositions, showed that their corresponding binding energies are slightly biased towards the stronger metal bonding interactions (i.e. Pt–Pt), in agreement with previous suggestions [18]. A subsequent in-depth study of 34-atom clusters, for all compositions, showed five structural motifs (as identified in a previous study using a simple averaged potential [10]) and a complex inter-dependence of chemical ordering and structure, with three main types of chemical ordering being observed: core–shell; spherical cap; and ball-and-cup (intermediate between the first two types). The core–shell configuration is generally found for potential weighting factors in the range $0 \leq w \leq 0.6$ (i.e. from strongly Pt-biased to slightly Pd-biased), all of which have negative excess energies corresponding to energetically favourable exothermic mixing. This, coupled with the fact that, for all except very low core/cap atom concentrations, the core–shell configuration has a significantly greater interfacial area than the layer-like spherical cap configuration, leads to the conclusion that the term "segregated" may be misleading when applied to core–shell clusters, since the relatively high interfacial area ensures a high number of heteronuclear (e.g. Pd–Pt) bonds. As a further example of this concept, the core–shell "segregated" pIh isomer of $Ag_{32}Au_6$, which has a hexagonal ring of six Au atoms at the centre of the 6-fold pancake structure, has almost the maximum number of Ag–Au bonds possible for this composition [20].

In this study, we found that average ($w = 0.5$) or slightly Pd-biased weighting factors ($w = 0.6$) stabilise the Dh–cp(DT) structural motif, previously reported as the DFT global minimum in the region of most exothermic mixing for 34-atom Pd–Pt clusters [10]. Thus, our results show that by finely tuning the Gupta potential, one can qualitatively reproduce structural and chemical ordering patterns observed at higher levels of theory (e.g. DFT).

It should be noted that the system (Pd–Pt) that we have studied here is one where there is a very small difference in lattice spacing, allowing us to make the approximation (see Sect. 2.5) that the atomic densities of the two metals are equal (actual atomic densities are: $6.8 \times 10^{28} \, m^{-3}$ (Pd) and $6.62 \times 10^{28} \, m^{-3}$ (Pt) [21]). The interplay of atomic size-mismatch and the strength of the heteronuclear interactions in determining the segregation properties of bimetallic nanoalloys [19, 22] is an interesting topic for future study.

Finally, although we hope that the results presented here will be helpful in future studies of catalysis on bimetallic particles, we do not suggest that 34-atom PdPt nanoparticles themselves are good candidates for such catalysts. We have chosen this size for the present investigation because our previous detailed Gupta-DFT study indicated a rich diversity of structural motifs for a computationally tractable cluster size [10] and we wanted to investigate how changes in the potential affect the atomic segregation, as well as the structures of nanoalloys. In order to study real catalysts under operational conditions, it would be necessary to investigate their thermal stability (with respect to both structure and chemical ordering). It is likely that the thermal energies ($k_B T$) corresponding to temperatures reached in catalytic converters are high enough to span a large number of permutational isomers (homotops) and structural isomers, but a detailed study of this requires calculation of activation barriers to isomerisation. Based on the energy spread of geometrical and permutational isomers observed in previous studies combining DFT and Gupta potential calculations [10, 23], we believe that it is likely that compact cluster structures will prevail at finite temperatures consistent with catalyst operation, though there may be a greater degree of Pd–Pt mixing. It should be noted that molecular dynamics simulations by Calvo, using the averaged Pd–Pt Gupta parameters, have indeed shown that there is some Pd–Pt mixing in the cores of onion-like shell clusters, even at relatively low temperatures, though Pd enrichment of the surface is maintained [5]. Other factors which must be considered in a realistic computational study of catalysis include the cluster–substrate interaction [24] and interactions with the reactant gases, which may induce cluster surface restructuring [25] and affect atomic segregation [26].

References

1. F. Cleri, V. Rosato, Phys. Rev. B **48**, 22 (1993)
2. R.L. Johnston, Dalton Trans. **22**, 4193 (2003)
3. L.O. Paz-Borbón, T.V. Mortimer-Jones, R.L. Johnston, A. Posada-Amarillas, G. Barcaro, A. Fortunelli, Phys. Chem. Chem. Phys. **9**, 5202 (2007)
4. D. Cheng, W. Wang, S. Huang, J. Phys. Chem. B **11**, 16193 (2006)
5. F. Calvo, Faraday Discuss. **138**, 75 (2008)
6. C. Massen, T.V. Mortimer-Jones, R.L. Johnston, J. Chem. Soc. Dalton Trans. **23**, 4375 (2002)
7. D. Bazin, D. Guillaume, C.H. Pichon, D. Uzio, S. Lopez, Oil Gas. Sci. Tech. Rev. IFP **60**, 801 (2005)
8. A.J. Renouprez, J.L. Rousset, A.M. Cadrot, Y. Soldo, L. Stievano, J. Alloy Comp. **328**, 50 (2001)
9. J.L. Rousset, L. Stievano, F.J. Cadete Santos Aires, C. Geantet, A.J. Renouprez, M. Pellarin, J. Catal. **202**, 163 (2001)
10. L.O. Paz-Borbón, R.L. Johnston, G. Barcaro, A. Fortunelli, J. Phys. Chem. C **111**, 2936 (2007)
11. E.M. Fernández, L.C. Balbás, L.A. Pérez, K. Michaelian, I.L. Garzón, Int. J. Mod. Phys. **19**, 2339 (2005)
12. L.D. Lloyd, R.L. Johnston, S. Salhi, N.T. Wilson, J. Mater. Chem. **14**, 1691 (2004)
13. L.D. Lloyd, R.L. Johnston, S. Salhi, J. Comp. Chem. **26**, 1069 (2005)
14. G. Rossi, R. Ferrando, A. Rapallo, A. Fortunelli, B.C. Curley, L.D. Lloyd, R.L. Johnston, J. Phys. Chem. **122**, 194309 (2005)
15. J. Wang, G. Wang, X. Chen, W. Lu, J. Zhao, Phys. Rev. B **66**, 014419 (2002)
16. F. Baletto, C. Mottet, R. Ferrando, Phys. Rev. B **66**, 155420 (2002)
17. L.O. Paz-Borbón, A. Gupta, R.L. Johnston, J. Mater. Chem. **18**, 4154 (2008)
18. F. Baletto, R. Ferrando, Rev. Mod. Phys. **77**, 371 (2005)
19. A. Christensen, P. Stoltze, J.K. Norskov, J. Phys. Condens. Matter **7**, 1047 (1995)
20. B.C. Curley, R.L. Johnston, G. Rossi, R. Ferrando, Eur. Phys. J. D **43**, 53 (2007)
21. C. Kittel, *Introduction to Solid State Physics*, 8th edn. (Wiley, USA, 2005)
22. A. Rapallo, G. Rossi, R. Ferrando, A. Fortunelli, B.C. Curley, L.D. Lloyd, G.M. Tarbuck, R.L. Johnston, J. Chem. Phys. **122**, 194308 (2005)
23. L.O. Paz-Borbón, R.L. Johnston, G. Barcaro, A. Fortunelli, J. Chem. Phys. **128**, 134517 (2008)
24. G. Barcaro, A. Fortunelli, Faraday Disscus. **138**, 37 (2008)
25. K. McKenna, A.L. Schluger, J. Phys. Chem. C **111**, 18848 (2007)
26. A.A. Herzing, M. Watanabe, J.K. Edwards, M. Conte, Z.-R. Tang, G.J. Hutchings, C.J. Kiely, Faraday Discuss. **138**, 337 (2008)

Chapter 7
Theoretical Study of Pd–Au Clusters

7.1 Introduction

Surface segregation in nanoalloys may be simply predicted on the basis of simple elemental properties such as cohesive energy (E_{coh}), surface energy (E_{surf}), atomic radius (r_a) and electronegativity (χ) [1]. These quantities are listed for Pd and Au in Table 7.1.

Preferential segregation of Au atoms to the surface of Pd–Au nanoalloys can be rationalised in terms of the marginally larger cohesive energy of Pd (favouring a Pd core, maximising Pd–Pd bonds) and the smaller surface energy of Au (atoms with lower surface energy forming a surface shell lower the cluster surface energy). The smaller atomic radius of Pd also leads to preferential core location of Pd atoms, as it helps to minimize bulk elastic strain. Finally, the slightly higher electronegativity of Au compared with Pd will lead to a certain degree of Pd to Au electron transfer, as found in DFT calculations by Yuan et al. [2]. This ionic contribution may be expected to favour Pd–Au mixing, although we have previously shown for Ag–Au clusters that metal (M) M-Au charge transfer can also favour surface enrichment by the more negatively charged atoms (Au) [3, 4]. Of course, the arguments presented above are rather simple and the fine details of cluster structure, chemical ordering (segregation or mixing) and surface site preferences depend critically on electronic structure. For example, recent DFT calculations by Yuan et al. on fcc-type cuboctahedral Pd–Au nanoparticles, have indicated that surface Pd atoms occupy (111), rather than (100) facets, thereby maximizing the number of relatively strong surface Pd–Au bonds [5]. In fact the ionic contribution to Pd–Au bonding has been used to predict that Pd–Au bonds are stronger than both Pd–Pd and Au–Au bonds [2]. These findings are consistent with earlier studies by Goodman and co-workers [6] and by Jose Yacaman and colleagues [7], who have also combined TEM and HAADF-STEM measurements with image simulation and classical molecular dynamics simulations, to study shell structure and defects in Pd–Au nanoalloys [7–10].

L. O. Paz Borbón, *Computational Studies of Transition Metal Nanoalloys,*
Springer Theses, DOI: 10.1007/978-3-642-18012-5_7,
© Springer-Verlag Berlin Heidelberg 2011

Table 7.1 Some elemental properties of Pd and Au

Metal	$E_{\text{coh}}/\text{eV(atom)}^{-1}$	$E_{\text{surface}}/\text{meV\,Å}^{-2}$	$r_{\text{a}}/\text{Å}$	χ
Pd	3.89	131.0	1.38	2.2
Au	3.81	96.8	1.44	2.4

In this sense, we have performed global optimizations on the size range ($N = 2$–50) and specific 98-atom size for Pd–Au clusters, with the interatomic interactions modeled by the Gupta many-body empirical potential (with three different potential parameterisations being investigated). DFT local-relaxations have been performed on all the "putative" GM structures for 1:1 Pd–Au compositions up to 50 atoms, for all sets of parameters. For nuclearities $N = 34, 38$ and 98 the structures and chemical ordering have been investigated as a function of composition at both the Gupta and DFT levels.

7.2 Computational Details

7.2.1 Genetic Algorithm

The GA parameters adopted in this work (for the three sets of parameters) were: population size = 40 clusters, crossover rate = 0.8, crossover type = 1-point weighted, selection = roulette, mutation rate = 0.1, mutation type = mutate-move, number of generations = 400, and number of GA runs for each composition = 100. DFT local-relaxation calculations were performed using NWChem5.1, while for specific 98 atom size clusters (LT T_d structures), we have used the shell optimization routine (see more specific details in Chap. 2).

7.2.2 Pd–Au Parameters

Three sets of potential parameters have been adopted in this work (see Table 7.2): (a) on in which the heteronuclear Pd–Au parameters are obtained as averages of the pure Pd–Pd and Au–Au parameters, hereafter referred to as "average" [3]; (b) one in which Pd–Pd, Pd–Au and Au–Au parameters were simultaneously fitted to experimental properties of bulk Pd, Au and Pd–Au alloys, hereafter referred to as "exp-fit", and (c) one in which the parameters for Pd–Au heteronuclear interactions were fitted to DFT calculations, hereafter referred to as "DFT-fit". The choice of set (a), in which the Pd–Au parameters have been rounded to the precision where the arithmetic and geometric means are the same, was based on our earlier work on Pd–Pt nanoalloys, where averaged parameters were found to qualitatively reproduce the experimental Pd surface segregation [11]. Recently, we

Table 7.2 Gupta potential parameters used in this study

Parameters	(a) Average (average)			(b) Experimental fitted parameters (exp-fit)			(c) DFT fitted parameters (DFT-fit)		
	Pd–Pd	Au–Au	Pd–Au	Pd–Pd	Au–Au	Pd–Au	Pd–Pd	Au–Au	Pd–Au
A (eV)	0.1746	0.2061	0.1900	0.1715	0.2096	0.2764	0.1653	0.2091	0.1843
ζ (eV)	1.7180	1.7900	1.7500	1.7019	1.8153	2.0820	1.6805	1.8097	1.7867
p	10.867	10.229	10.540	11.000	10.139	10.569	10.8535	10.2437	10.5420
q	3.7420	4.0360	3.8900	3.7940	4.0330	3.9130	3.7516	4.0445	3.8826
r_0(Å)	2.7485	2.8840	2.8160	2.7485	2.8840	2.8160	2.7485	2.8840	2.8160

have shown that the average Pd–Pt Gupta potential also gives results which are in reasonable agreement with DFT calculations [3, 12].

In parameter set (b), the mixed parameters A and σ have been fitted to the dissolution energies of one impurity Au atom into bulk Pd and vice versa. There values are derived from the enthalpy curves in the Pd–Au phase-diagram [13]. The dissolution energies are taken as the slope of the mixing enthalpy curve on each side of the phase-diagram—i.e. the Pd-rich phase for the dissolution energy of Au and the Au-rich phase for the dissolution energy of Pd. As the difference in size between the two elements is not negligible, we took into account possible relaxations around the impurity in our fitting procedure. The Pd–Au bulk phase-diagram presents a complete series of solid solutions with some possible ordered $L1_2$ AuCu$_3$ type phases near the composition Au$_{60}$Pd$_{40}$. This means that there is a strong tendency to mix the two elements in the bulk. It is interesting to note that exp-fit parameters set (b) has a pair (repulsive) energy scaling parameter (A) which is larger for Pd–Au than for either Pd–Pd or Au–Au: in our previous work, a high heteronuclear A value was found to lead to layer-like segregation [11]. However, set (b) also has a many-body (attractive) energy scaling parameter (σ) which is greatest for Pd–Au: in our previous work, this was found to lead to heteronuclear mixing [11]. As it is shown below, for set (b) the many-body term wins out, so that, overall, the fitted potential favours Pd–Au mixing, as expected from the fitting procedure detailed above. The pair and many-body range exponents (p and q) for Pd–Au interactions (and the atomic radii r_a) were taken as arithmetic means for the pure metal values.

In parameter set (c), the parameters have been fitted to the results of first-principles DFT calculations. This is consists of taking the DFT cohesive energy curves of the pure systems and rescaling these to fit the experimental ones (to obtain the exact values of the cohesive energy, lattice parameter and stickiness). The obtained rescaling factors are used to rescale the cohesive energy curves of the systems given by ordered intermetallic phases, which are then used to fit the parameters of the heteronuclear Pd–Au interaction. For the metals A and B (i.e. Pd and Au), the chosen ordered alloys are of the type A_nB_m, with $m + n = 4$: in the case $m = n = 2$, we chose the $L1_0$ ordering of the alloy, whereas in the case $m = 1$ and $n = 3$ and vice versa we chose the $L1_2$ ordering.

7.3 Results and Discussion

7.3.1 Gupta Potential Calculations on (Pd–Au)$_{N/2}$ Clusters (N = 2–50)

Some of the putative GM for (Pd–Au)$_{N/2}$ clusters (with $N = 2-50$) are compared in Fig. 7.1 for the average, exp-fit and DFT-fit parameters. The symmetries and total energies of all the GM in the Gupta level (in this size range) are listed in Tables 7.3, 7.4 and 7.5 (for the average, exp-fit and DFT-fit parameters, respectively). The geometrical structures found for the three sets of parameters are often the same, but generally exhibit different chemical ordering. From the structures shown in Fig. 7.1, it is clear that the average potential favours Pd$_{core}$Au$_{shell}$ segregation (as found in our previous study [3]), while the exp-fit parameters favours more mixed configurations, with more Pd–Au bonds. The DFT-fit parameters predicts a degree of Au/Pd mixing between the two other potentials. A wide range of structural motifs are found after performing GA global optimizations for the three sets of parameters. Structural families includes icosahedra (e.g. $N = 54$), Marks decahedra (e.g. $N = 98$) and fcc-type structures (e.g. the truncated octahedron, $N = 38$) [1, 3, 14–17].

Figure 7.2a shows the calculated Gupta potential binding energies (E_b^{Gupta}) for 1:1 Pd–Au nanoalloys with 2–50 atoms. In each case, the binding energy is

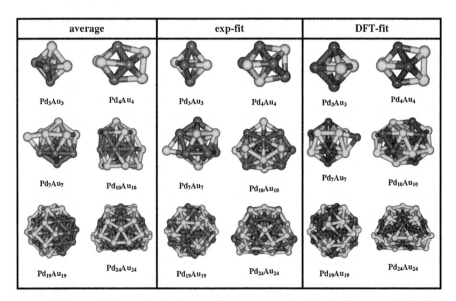

Fig. 7.1 Comparison of selected GM found for (Pd--Au)$_{N/2}$ clusters with ($N = 2$–50), for the average, exp-fit and DFT-fit parameters. Au and Pd atoms are denoted by *light* and *dark colours*, respectively

Table 7.3 Total cluster energies (V^{clus}) and symmetries for GM of $(Pd-Au)_{N/2}$ from Gupta potential calculations (average parameters)

Composition	Symmetry	V_{clus}/eV	Composition	Symmetry	V_{clus}/eV	Composition	Symmetry	V_{clus}/eV
Pd_1Au_1	C_∞	-4.51746	$Pd_{18}Au_{18}$	C_1	-124.522	$Pd_{35}Au_{35}$	C_1	-248.801
Pd_2Au_2	C_{2v}	-11.3834	$Pd_{19}Au_{19}$	C_s	-131.937	$Pd_{36}Au_{36}$	C_1	-256.115
Pd_3Au_3	C_{3v}	-18.2640	$Pd_{20}Au_{20}$	C_s	-139.019	$Pd_{37}Au_{37}$	C_1	-263.573
Pd_4Au_4	C_{2v}	-24.9984	$Pd_{21}Au_{21}$	C_1	-146.223	$Pd_{38}Au_{38}$	C_1	-270.947
Pd_5Au_5	C_s	-31.9459	$Pd_{22}Au_{22}$	C_1	-153.430	$Pd_{39}Au_{39}$	C_1	-278.055
Pd_6Au_6	C_s	-38.8571	$Pd_{23}Au_{23}$	C_1	-160.861	$Pd_{40}Au_{40}$	C_1	-285.291
Pd_7Au_7	C_s	-45.9385	$Pd_{24}Au_{24}$	C_{2v}	-168.321	$Pd_{41}Au_{41}$	C_1	-292.484
Pd_8Au_8	C_1	-52.9995	$Pd_{25}Au_{25}$	C_1	-175.444	$Pd_{42}Au_{42}$	C_1	-299.771
Pd_9Au_9	C_1	-60.0126	$Pd_{26}Au_{26}$	C_1	-182.814	$Pd_{43}Au_{43}$	C_1	-307.145
$Pd_{10}Au_{10}$	C_1	-67.0545	$Pd_{27}Au_{27}$	C_s	-190.345	$Pd_{44}Au_{44}$	C_1	-314.485
$Pd_{11}Au_{11}$	C_s	-74.2458	$Pd_{28}Au_{28}$	C_1	-197.395	$Pd_{45}Au_{45}$	C_1	-322.001
$Pd_{12}Au_{12}$	C_1	-81.2728	$Pd_{29}Au_{29}$	C_1	-204.744	$Pd_{46}Au_{46}$	C_1	-329.356
$Pd_{13}Au_{13}$	C_1	-88.5106	$Pd_{30}Au_{30}$	C_1	-211.968	$Pd_{47}Au_{47}$	C_1	-336.796
$Pd_{14}Au_{14}$	C_s	-95.6624	$Pd_{31}Au_{31}$	C_1	-219.220	$Pd_{48}Au_{48}$	C_1	-344.160
$Pd_{15}Au_{15}$	C_1	-102.788	$Pd_{32}Au_{32}$	C_1	-226.758	$Pd_{49}Au_{49}$	C_1	-351.646
$Pd_{16}Au_{16}$	C_1	-109.991	$Pd_{33}Au_{33}$	C_1	-234.046	$Pd_{50}Au_{50}$	C_1	-359.365
$Pd_{17}Au_{17}$	C_s	-117.273	$Pd_{34}Au_{34}$	C_1	-241.281			

Table 7.4 Total cluster energies (V^{clus}) and symmetries for GM of $(Pd-Au)_{N/2}$ from Gupta potential calculations (exp-fit parameters)

Composition	Symmetry	V_{clus}/eV	Composition	Symmetry	V_{clus}/eV	Composition	Symmetry	V_{clus}/eV
Pd_1Au_1	C_∞	−4.79234	$Pd_{18}Au_{18}$	C_1	−128.811	$Pd_{35}Au_{35}$	C_1	−256.470
Pd_2Au_2	C_{2v}	−11.8711	$Pd_{19}Au_{19}$	C_1	−136.621	$Pd_{36}Au_{36}$	C_1	−263.963
Pd_3Au_3	C_{2v}	−19.0004	$Pd_{20}Au_{20}$	C_1	−143.918	$Pd_{37}Au_{37}$	C_1	−271.619
Pd_4Au_4	C_1	−26.0123	$Pd_{21}Au_{21}$	C_1	−151.252	$Pd_{38}Au_{38}$	C_1	−279.038
Pd_5Au_5	C_s	−33.2528	$Pd_{22}Au_{22}$	C_1	−158.775	$Pd_{39}Au_{39}$	C_1	−286.501
Pd_6Au_6	C_s	−40.3840	$Pd_{23}Au_{23}$	C_1	−166.474	$Pd_{40}Au_{40}$	C_1	−293.768
Pd_7Au_7	C_1	−47.7126	$Pd_{24}Au_{24}$	C_1	−173.861	$Pd_{41}Au_{41}$	C_1	−301.384
Pd_8Au_8	C_s	−55.0601	$Pd_{25}Au_{25}$	C_1	−181.424	$Pd_{42}Au_{42}$	C_1	−308.801
Pd_9Au_9	C_1	−62.2956	$Pd_{26}Au_{26}$	C_1	−189.133	$Pd_{43}Au_{43}$	C_1	−316.229
$Pd_{10}Au_{10}$	C_s	−69.5662	$Pd_{27}Au_{27}$	C_1	−196.835	$Pd_{44}Au_{44}$	C_1	−324.125
$Pd_{11}Au_{11}$	C_1	−77.0285	$Pd_{28}Au_{28}$	C_1	−204.145	$Pd_{45}Au_{45}$	C_1	−331.571
$Pd_{12}Au_{12}$	C_1	−84.2377	$Pd_{29}Au_{29}$	C_1	−211.666	$Pd_{46}Au_{46}$	C_1	−339.164
$Pd_{13}Au_{13}$	C_1	−91.7021	$Pd_{30}Au_{30}$	C_1	−219.133	$Pd_{47}Au_{47}$	C_1	−346.735
$Pd_{14}Au_{14}$	C_1	−99.1405	$Pd_{31}Au_{31}$	C_1	−226.555	$Pd_{48}Au_{48}$	C_1	−354.216
$Pd_{15}Au_{15}$	C_1	−106.459	$Pd_{32}Au_{32}$	C_1	−234.035	$Pd_{49}Au_{49}$	C_1	−361.886
$Pd_{16}Au_{16}$	D_{2d}	−113.972	$Pd_{33}Au_{33}$	C_1	−241.427	$Pd_{50}Au_{50}$	C_1	−369.647
$Pd_{17}Au_{17}$	C_1	−121.338	$Pd_{34}Au_{34}$	C_1	−248.727			

Table 7.5 Total cluster energies (V^{clus}) and symmetries for GM of $(Pd-Au)_{N/2}$ from Gupta potential calculations (DFT-fit parameters)

Composition	Symmetry	V_{clus}/eV	Composition	Symmetry	V_{clus}/eV	Composition	Symmetry	V_{clus}/eV
Pd_1Au_1	C_∞	−4.74175	$Pd_{18}Au_{18}$	C_1	−128.244	$Pd_{35}Au_{35}$	C_1	−255.520
Pd_2Au_2	C_{2v}	−11.7847	$Pd_{19}Au_{19}$	C_3	−135.904	$Pd_{36}Au_{36}$	C_1	−263.069
Pd_3Au_3	C_{2v}	−18.8930	$Pd_{20}Au_{20}$	C_1	−143.185	$Pd_{37}Au_{37}$	C_1	−270.585
Pd_4Au_4	D_{2d}	−25.8700	$Pd_{21}Au_{21}$	C_1	−150.474	$Pd_{38}Au_{38}$	C_1	−278.233
Pd_5Au_5	C_s	−33.0114	$Pd_{22}Au_{22}$	C_1	−158.042	$Pd_{39}Au_{39}$	C_1	−285.419
Pd_6Au_6	C_{5v}	−40.1703	$Pd_{23}Au_{23}$	C_1	−165.550	$Pd_{40}Au_{40}$	C_1	−292.735
Pd_7Au_7	C_s	−47.4721	$Pd_{24}Au_{24}$	C_1	−173.091	$Pd_{41}Au_{41}$	C_1	−300.231
Pd_8Au_8	C_1	−54.7912	$Pd_{25}Au_{25}$	C_s	−180.454	$Pd_{42}Au_{42}$	C_1	−307.741
Pd_9Au_9	C_1	−61.9610	$Pd_{26}Au_{26}$	C_1	−188.028	$Pd_{43}Au_{43}$	C_1	−315.127
$Pd_{10}Au_{10}$	D_2	−69.2190	$Pd_{27}Au_{27}$	C_2	−195.753	$Pd_{44}Au_{44}$	C_1	−322.613
$Pd_{11}Au_{11}$	C_1	−76.6674	$Pd_{28}Au_{28}$	C_1	−203.098	$Pd_{45}Au_{45}$	C_1	−330.381
$Pd_{12}Au_{12}$	C_1	−83.9422	$Pd_{29}Au_{29}$	C_1	−210.531	$Pd_{46}Au_{46}$	C_1	−337.793
$Pd_{13}Au_{13}$	C_3	−91.2541	$Pd_{30}Au_{30}$	C_1	−217.889	$Pd_{47}Au_{47}$	C_1	−345.553
$Pd_{14}Au_{14}$	C_1	−98.7046	$Pd_{31}Au_{31}$	C_1	−225.467	$Pd_{48}Au_{48}$	C_1	−352.973
$Pd_{15}Au_{15}$	C_s	−105.975	$Pd_{32}Au_{32}$	C_1	−232.996	$Pd_{49}Au_{49}$	C_1	−360.661
$Pd_{16}Au_{16}$	C_1	−113.384	$Pd_{33}Au_{33}$	C_1	−240.444	$Pd_{50}Au_{50}$	C_1	−368.684
$Pd_{17}Au_{17}$	C_s	−120.744	$Pd_{34}Au_{34}$	C_1	−247.881			

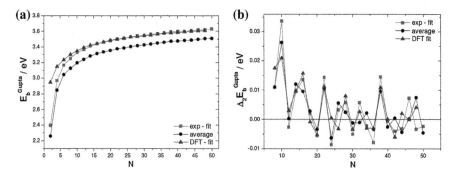

Fig. 7.2 Plots of **a** binding energies (E_b^{Gupta}) and **b** second differences in binding energy (ΔE_b^{Gupta}) for $(Pd-Au)_{N/2}$ clusters with ($N = 2$–50), at the Gupta level, calculated using the average (*circles*), exp-fit (*squares*) and DFT-fit (*triangles*) parameters

reported for the putative GM found by the BCGA. The binding energies (E_b^{Gupta}) of the GM obtained for the exp-fit parameters are higher than those obtained for the average parameters, but slightly higher than for the DFT-fit parameters, indicating that Pd–Au mixing is preferred (at the EP level) in this size regime. Figure 7.2b shows plots of the corresponding second differences in binding energy ($\Delta_2 E_b^{Gupta}$). Large peaks in $\Delta_2 E_b^{Gupta}$ correspond to structures which have high stability relative to their neighbours. However, for Pd–Au nanoalloys, the $\Delta_2 E_b^{Gupta}$ values are quite small (for the three sets of parameters) because of the relatively small difference in cohesive energy of Pd and Au (see Table 7.1). The $\Delta_2 E_b^{Gupta}$ plots are similar for the three sets of parameters (which tend to find the same structure as GM, with different chemical order), where the discrepancies are due to the different structures and homotops stabilised by these potentials.

7.3.2 DFT Calculations on $(Pd-Au)_{N/2}$ Clusters ($N = 2$–25)

DFT local-relaxations have been performed on all the putative GM structures for 1:1 Pd–Au clusters with up to 50 atoms, for the three sets of parameters. Figure 7.3a shows a plot of the binding energy (E_b^{DFT}) versus the total number of atoms (N). Here one can see that there is not much difference at the DFT level between the GM isomers found for the average, exp-fit and DFT-fit parameters. Figure 7.3b shows the differences in DFT binding energies between the fitted and average parameters (i.e. $\Delta E_b^{DFT} = E_b^{DFT}(\text{average}) - E_b^{DFT}(\text{fitted})$); highlighting that, for some specific sizes one on both fitted potentials may give rise to more stable structures (indicated by negative ΔE_b^{DFT} values) than the average parameters.

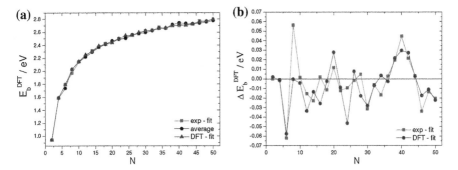

Fig. 7.3 **a** DFT binding energies (E_b^{DFT}) after local re-minimization of the putative GM for the average (*circles*), exp-fit (*squares*) and DFT-fit (*triangles*) Gupta parameters, for (Pd–Au)$_{N/2}$ clusters with $N = 2$–50. **b** Differences in DFT binding energies (ΔE_b^{DFT}): exp-fit and DFT-fit versus average parameters, where negative values indicate more stable clusters than those found using the average parameters

7.3.3 Structural Analysis

In Fig. 7.4, the total number of bonds (nearest neighbour contacts) of each type and the chemical ordering parameter (σ, see Eq. 2.33), for the three parameters, are plotted as a function of cluster size (N) for 1:1 Pd–Au clusters. For the average potential, one can see that the numbers of Pd–Pd and Pd–Au bonds increase monotonically (quasi-linearly) with N and are nearly equal; see Fig. 7.4a. The number of Au–Au bonds also increases with N, but less steeply than for N_{Pd-Pd} and N_{Pd-Au}. For both fitted parameters (Fig. 7.4b and c), the rate of increase in the number of bonds is in the order Pd–Au > Pd–Pd > Au–Au, indicating the greater Pd–Au mixing favoured by fitted potentials. The corresponding σ values for all three sets of parameters are plotted in Fig. 7.4d. We find that, for the average parameters, σ is positive for cluster sizes $N > 10$, corresponding to Pd$_{core}$Au$_{shell}$ segregation. In contrast, for both fitted parameters, the σ values are negative for all cluster sizes, indicating that the clusters have significant Pd–Au mixing ($N_{Pd-Au} > N_{Pd-Pd} + N_{Au-Au}$).

7.4 34-Atom Pd–Au Clusters

Figure 7.5 shows the excess energy (Δ_{34}^{Gupta}) calculated at the Gupta level (for the three sets of parameters) for all compositions of Pd$_{34-m}$Au$_m$ clusters, plotted against the number of Au atoms (m). We selected 34-atom clusters in order to allow comparisons to be made with the results of our previous in-depth study of 34-atom Pd–Pt clusters [12, 18]. The tendency of the exp-fit and DFT-fit parameters to favour Pd–Au mixing is emphasized by the significantly more negative excess energies obtained for these potentials. These large negative excess energies

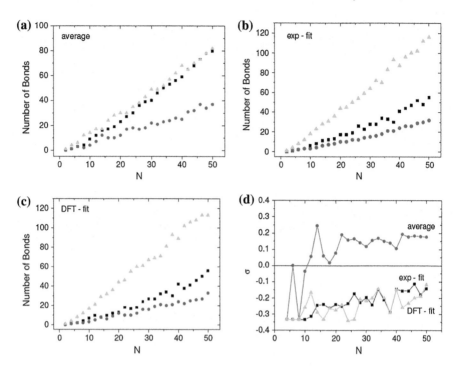

Fig. 7.4 a, b, c Numbers of Pd–Pd (*black*), Au–Au (*red*) and Pd–Au (*green*) nearest neighbour bonds as a function of nuclearity ($N = 2$–50) for 1:1 composition Pd–Au clusters, for GM obtained using the average, exp-fit and DFT-fit parameters. **d** Comparison of the chemical order parameter (σ) for the average (*red*), exp-fit (*black*) and DFT-fit (*green*)

(i.e. representing highly exothermic mixing) are reminiscent of the excess energy curves obtained for 34-atom Pd–Pt clusters when weighting the heteronuclear Pd–Pt interactions more towards the strongest (Pt–Pt) homonuclear interactions [12].

The excess energies (Δ_{34}^{DFT}) calculated at the DFT level (based on the GM found for the three sets of potentials) are shown in Fig. 7.6 for the composition range $Pd_{24}Au_{10}$ to Pd_6Au_{29}, which corresponds to the region of most negative excess energies for these potentials at the Gupta level; see Fig. 7.7 and Table 7.6 for a comparative analysis of GM structures for the three sets of potentials, at the DFT level. Compared with the relatively smooth Δ_{34}^{Gupta} curves, Δ_{34}^{DFT} plots are rather jagged, especially for both exp-fit and DFT-fit parameter curves. It is not clear what the preference is, at the DFT level, for structures based on these potentials; though the exp-fit potential gives marginal lower excess energies structures, compared to the average and DFT-fit parameters GM. The jagged Δ_{34}^{DFT} plots are reminiscent of those obtained for 34-atom Pd–Pt clusters and may (as in the Pd–Pt case) indicate that there are differences between the energy ordering of structural motifs at the Gupta and DFT levels [18]. The reason is that some of the motifs that are energetically close at the Gupta level are disfavored by DFT, so that Δ_{34}^{DFT}

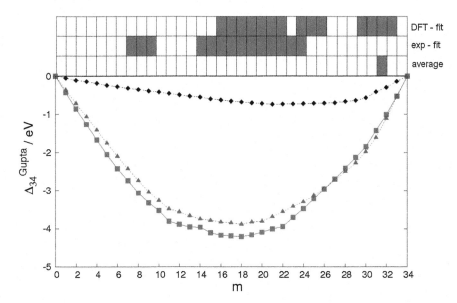

Fig. 7.5 (*Bottom*) Gupta potential excess energies ($\Delta_{34}^{\text{Gupta}}$) for Pd$_{34-m}Au_m$ nanoalloys for the average (*black line*), exp-fit (*red line*) and DFT-fit (*blue line*) parameters. (*Top*) *Dark blue squares* indicate poly-icosahedral structures; *red squares* pIh-6 ("pancake-type") structures; *white squares* denote incomplete decahedral structures

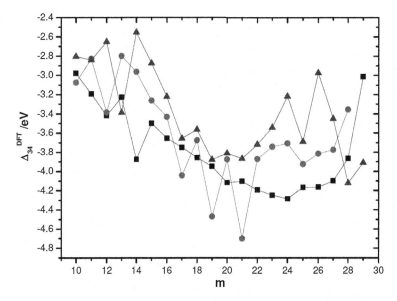

Fig. 7.6 DFT excess energies (Δ_{34}^{DFT}) after local reminimization of the putative GM for the average (*squares*), exp-fit (*circles*) and DFT-fit (*triangles*) Gupta parameters, for Pd$_{34-m}$Au$_m$ nanoalloys in the range $m = 10$–29

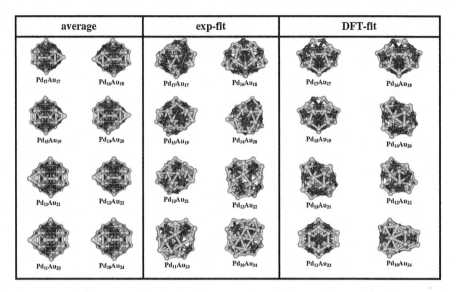

average		exp-fit		DFT-fit	
$Pd_{17}Au_{17}$	$Pd_{16}Au_{18}$	$Pd_{17}Au_{17}$	$Pd_{16}Au_{18}$	$Pd_{17}Au_{17}$	$Pd_{16}Au_{18}$
$Pd_{15}Au_{19}$	$Pd_{14}Au_{20}$	$Pd_{15}Au_{19}$	$Pd_{14}Au_{20}$	$Pd_{15}Au_{19}$	$Pd_{14}Au_{20}$
$Pd_{13}Au_{21}$	$Pd_{12}Au_{22}$	$Pd_{13}Au_{21}$	$Pd_{12}Au_{22}$	$Pd_{13}Au_{21}$	$Pd_{12}Au_{22}$
$Pd_{11}Au_{23}$	$Pd_{10}Au_{24}$	$Pd_{11}Au_{23}$	$Pd_{10}Au_{24}$	$Pd_{11}Au_{23}$	$Pd_{10}Au_{24}$

Fig. 7.7 GM structures obtained after DFT optimization (for the three sets of Gupta parameters) for $Pd_{34-m}Au_m$ clusters in the composition range $Pd_{17}Au_{17}$–$Pd_{10}Au_{24}$

curve obtained by reoptimizing the Gupta GM at the DFT level becomes irregular whenever such motifs are predicted as the Gupta GM.

The structures of the average, exp-fit and DFT-fit GM (after DFT reminimization) are shown in Fig. 7.7, for the composition range $Pd_{17}Au_{17}$ to $Pd_{10}Au_{24}$. This composition range includes the most negative Δ_{34}^{DFT} values. In this region, all of the putative GM for the average parameters have incomplete decahedral structures (but no Dh–cp(DT) structures as found previously for 34-atom Pd–Pt clusters). This probably explains the rather smooth curve for the average parameter sets (red line, see Fig. 7.6) with the Δ_{34}^{DFT} minimum at composition $Pd_{10}Au_{24}$. On the other hand, the structures derived from both fitted parameters tend to maximize the number of Pd–Au bonds; hence, they tend to form poly-icosahedral arrangements, in this composition range. It is interesting to note that the lowest Δ_{34}^{DFT} compositions correspond to two structures obtained using exp-fit parameters (i.e. compositions $Pd_{13}Au_{21}$ and $Pd_{15}Au_{19}$). When comparing these two structures, with those found using the DFT-fit parameters, we notice that they have a distorted fivefold shape, compared to the more asymmetric (amorphous) structures found using the DFT-fit parameters.

7.5 38-Atom Pd–Au Clusters

Figure 7.8 shows the excess energy (Δ_{38}^{Gupta}) calculated at the Gupta level (for the three sets of potentials) for all compositions of $Pd_{38-m}Au_m$ clusters, plotted against the number of Au atoms (m). We selected 38-atom clusters because size 38 is a

Table 7.6 DFT excess energy (Δ_{34}^{DFT}/eV), structures and symmetries for GM of $Pd_{34-m}Au_m$ clusters (for the three sets of parameters), in the $m = 10-29$ composition range

Composition	(a) Average			(b) exp-fit			(c) DFT-fit		
	Struct.	Symm.	Δ_{34}^{DFT}/eV	Struct.	Symm.	Δ_{34}^{DFT}/eV	Struct.	Symm.	Δ_{34}^{DFT}/eV
$Pd_{24}Au_{10}$	Inc. Dh	C_1	−2.97699	Inc. Dh	C_1	−3.07223	Inc. Dh	C_1	−2.80414
$Pd_{23}Au_{11}$	Inc. Dh	C_1	−3.19469	Inc. Dh	C_3	−2.82733	Inc. Dh	C_1	−2.84063
$Pd_{22}Au_{12}$	Inc. Dh	C_1	−3.41783	Inc. Dh	C_1	−3.38517	Inc. Dh	C_2	−2.64947
$Pd_{21}Au_{13}$	Inc. Dh	C_s	−3.22734	Inc. Dh	C_1	−2.79739	Inc. Dh	C_1	−3.38667
$Pd_{20}Au_{14}$	Inc. Dh	C_1	−3.87499	pIh	C_1	−2.96339	Inc. Dh	C_1	−2.55399
$Pd_{19}Au_{15}$	Inc. Dh	C_s	−3.49674	pIh	C_1	−3.26272	Inc. Dh	C_1	−2.87404
$Pd_{18}Au_{16}$	Inc. Dh	C_1	−3.65185	pIh	C_1	−3.43143	pIh	C_1	−3.22233
$Pd_{17}Au_{17}$	Inc. Dh	C_s	−3.74709	pIh	C_s	−4.04098	pIh	C_s	−3.65516
$Pd_{16}Au_{18}$	Inc. Dh	C_1	−3.85594	pIh	C_1	−3.67362	pIh	C_s	−3.56327
$Pd_{15}Au_{19}$	Inc. Dh	D_s	−3.94846	pIh	C_2	−4.46821	pIh	C_s	−3.87802
$Pd_{14}Au_{20}$	Inc. Dh	C_1	−3.94846	pIh	C_1	−3.87499	pIh	C_s	−3.81047
$Pd_{13}Au_{21}$	Inc. Dh	C_s	−4.11718	pIh	C_1	−4.69951	pIh	C_1	−3.86769
$Pd_{12}Au_{22}$	Inc. Dh	C_s	−4.10357	pIh	C_1	−3.87227	Inc. Dh	C_1	−3.71991
$Pd_{11}Au_{23}$	Inc. Dh	C_1	−4.19337	pIh	C_1	−3.74165	pIh	C_1	−3.54194
$Pd_{10}Au_{24}$	Inc. Dh	C_s	−4.24779	pIh	C_1	−3.70627	pIh	C_s	−3.21724
Pd_9Au_{25}	Inc. Dh	C_s	−4.28589	Inc. Dh	C_1	−3.92397	pIh	C_s	−3.69086
Pd_8Au_{26}	Inc. Dh	C_s	−4.16344	Inc. Dh	C_1	−3.81512	pIh-6	D_{2h}	−2.97306
Pd_7Au_{27}	Inc. Dh	C_s	−4.09813	Inc. Dh	C_1	−3.7743	pIh	C_s	−3.45032
Pd_6Au_{28}	Inc. Dh	C_1	−3.86410	Inc. Dh	C_1	−3.35524	pIh	C_1	−4.12119
Pd_5Au_{29}	Inc. Dh	C_s	−3.01237	—	—	—	pIh	C_s	−3.90922

Dashed lines indicate no SCF convergence

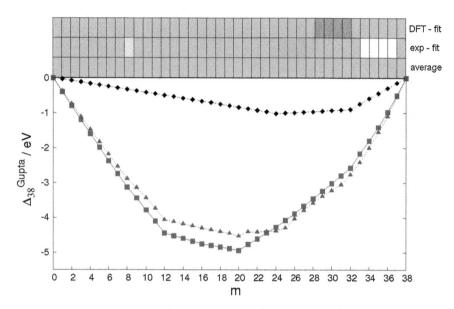

Fig. 7.8 (*Bottom*) Gupta potential excess energies ($\Delta_{38}^{\text{Gupta}}$) for $Pd_{34-m}Au_m$ nanoalloys for the average (*black line*), exp-fit (*red line*) and DFT-fit (*blue line*) parameters. (*Top*) *Light blue squares* denotes compositions at which TO structures were found; *yellow* denotes Oh-Ih (closed-packed) structures; *green* denotes Ih-Mackay (pseudo-fivefold structures); *orange* denotes pIh-5 (fivefold- "pancakes"), while *white* denotes incomplete decahedral structures

"magic" number for the truncated octahedral (TO) structure, which has fcc packing (the bulk crystalline structure for both Pd and Au), and also for the poly-icosahedral, pIh-6 "six-fold pancake" structure. These structures, along with other structural motifs, have been studied in previous work on 38-atom nanoalloy clusters (see Chap. 5 as well) [3, 15, 16].

The tendency of the exp-fit parameters to favour Pd–Au mixing is again manifest in the much more negative excess energies ($\Delta_{38}^{\text{Gupta}}$) obtained for this potential, which is also found for the DFT-fit $\Delta_{38}^{\text{Gupta}}$ values (see Fig. 7.8). Less negative $\Delta_{38}^{\text{Gupta}}$ values are found for the average parameters, for which the lowest $\Delta_{38}^{\text{Gupta}}$ value corresponds to the Au-rich composition $Pd_{14}Au_{24}$. As shown in Fig. 7.9 (which reproduces the structures found as the putative DFT-GM for the average, exp-fit and DFT-fit parameters in the composition range $Pd_{19}Au_{19}$ to $Pd_{13}Au_{25}$) this structure has regular O_h symmetry, consisting of an fcc-like trun-cated octahedron (TO) with an octahedral Pd_6 core surrounded by a shell of 24 Au atoms, with 8 Pd atoms occupying the centres of the (111) facets. The corre-sponding minimum in $\Delta_{38}^{\text{Gupta}}$ for the exp-fit and DFT-fit parameters is at $Pd_{18}Au_{20}$, which is consistent with these parameters favouring Pd–Au mixing—which should be maximized around the 1:1 composition (although we still see a preference of Pd

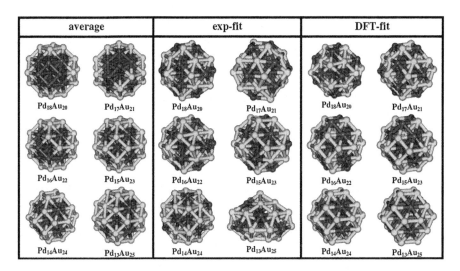

average	exp-fit	DFT-fit

Fig. 7.9 GM structures (for the three sets of Gupta parameters) for $Pd_{38-m}Au_m$ clusters in the composition range $Pd_{18}Au_{20}$–$Pd_{13}Au_{25}$

atoms to occupy core positions in the cluster, for the three sets of potentials; i.e. $Pd_{core}Au_{shell}$ segregation). The preferred sites for occupation by Pd and Au atoms are evident from the shapes of the Δ_{38}^{Gupta} curves. Specifically, for Δ_{38}^{Gupta} (average) there are straight sections corresponding to: the first 24 Au atoms occupying surface (100) TO sites (square faces); the next 8 atoms occupying the centres of the surface (111) facets (hexagonal faces); the final 6 atoms occupying the octahedral core of the cluster. For Δ_{38}^{Gupta} (exp-fit), however, the shape of the curve is different: the first 12 Au atoms occupy half of the (100) sites; the next 8 occupy the (111) sites; the next 12 occupy the remainder of the (100) sites; and the final 6 occupy the octahedral core.

For the average parameters (see Fig. 7.9), the GM structures obtained for all compositions are TO with surface Au enrichment (i.e. idealized $Pd_{core}Au_{shell}$ configurations), in agreement with our previous studies [3]. In agreement with the DFT calculations [2], and as discussed above for $Pd_{14}Au_{24}$, surface Pd atoms preferentially occupy (111) facets. For the exp-fit parameters (see Fig. 7.9), the GM for most compositions are also TO, though there is a five-fold symmetry structure found as GM at composition $Pd_{13}Au_{25}$ (an icosahedral Mackay type structure). Again, the exp-fit parameters show significantly more Pd–Au mixing, though they tend to favour surface Pd atoms occupying (100) rather than (111) sites, in contrast to the findings of Yuan et al. [2]. This might be due to an incorrect interplay between metal binding and equilibrium distance.

The excess energies (Δ_{38}^{DFT}) calculated at the DFT level (based on the GM found for all three parameters) are shown in Fig. 7.10 for the composition range $Pd_{18}Au_{20}$–$Pd_{13}Au_{25}$. The point group symmetries, structures types and Δ_{38}^{DFT}

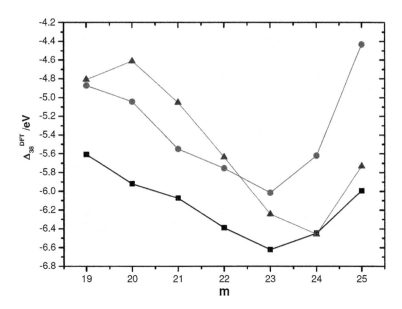

Fig. 7.10 DFT excess energies (Δ_{38}^{DFT}) after local re-minimization of the putative GM for the average (*squares*), exp-fit (*circles*) and DFT-fit (*triangles*) Gupta parameters, for $Pd_{38-m}Au_m$ nanoalloys, in the range $m = 19-25$

energies are listed in Table 7.7. In contrast to the plots of Δ_{34}^{Gupta} shown in Fig. 7.8, the Δ_{38}^{DFT} plots are quite smooth and it is also clear that in this composition range the isomers predicted to be the GM by the average parameters have consistently more negative excess energies after DFT reminimization than those from the exp-fit and DFT-fit parameters. The smoothness of the plots and lack of fluctuations is probably because all of the structures (with the above mentioned exception of $Pd_{13}Au_{25}$ (exp-fit)) have the same geometric structure (TO), as shown in Fig. 7.9. It should be noted that the minimum in the Δ_{38}^{DFT} plots for the average and exp-fit parameters occurs at composition $Pd_{15}Au_{23}$—i.e. one less Au atom than the minimum of Δ_{38}^{Gupta} (average) and three more Au atoms than the minimum of Δ_{38}^{Gupta} (exp-fit). Thus, for 38-atom Pd–Au clusters in this composition range we can say that the average parameters produce TO homotops which are closer to the true DFT GM (for this structure type) than those found using the exp-fit parameters.

It is interesting to note that the DFT-fit parameters also predict TO structures as GM, with a higher mixing of Pd–Au bonds, compared to the GM structures predicted by the average parameters. The DFT-fit parameters Δ_{38}^{Gupta} curve is rather smooth, similar to the other two Δ_{38}^{Gupta} curves, with a minimum at composition $Pd_{14}Au_{24}$ (a highly-symmetrical O_h structure), though it also has a bond maximum at $Pd_{18}Au_{20}$. The increased mixing on going from the average to the exp-fit parameters has been confirmed by a detailed analysis (for 38-atom Pd–Au clusters)

Table 7.7 DFT excess energy (Δ_{38}^{DFT}/eV), structures and symmetries for GM of $Pd_{38-m}Au_m$ clusters (for the three sets of parameters), in the $m = 19$–25 composition range

Composition	(a) Average			(b) exp-fit			(c) DFT-fit		
	Struct.	Symm.	Δ_{38}^{DFT}/eV	Struct.	Symm.	Δ_{38}^{DFT}/eV	Struct.	Symm.	Δ_{38}^{DFT}/eV
$Pd_{19}Au_{19}$	TO	C_s	−5.60839	TO	C_3	−4.87095	TO	C_3	−4.80752
$Pd_{18}Au_{20}$	TO	C_{4v}	−5.92133	TO	O_h	−5.04238	TO	C_1	−4.61148
$Pd_{17}Au_{21}$	TO	C_1	−6.07372	TO	C_s	−5.54853	TO	C_s	−5.05240
$Pd_{16}Au_{22}$	TO	C_s	−6.38938	TO	C_3	−5.75262	TO	D_{2d}	−5.63511
$Pd_{15}Au_{23}$	TO	C_s	−6.62068	TO	C_3	−6.01385	TO	C_{2v}	−6.24308
$Pd_{14}Au_{24}$	TO	O_h	−6.44652	TO	C_3	−5.61928	TO	O_h	−6.45662
$Pd_{13}Au_{25}$	TO	C_{3v}	−5.99208	pIh-5	C_1	−4.43283	TO	C_{4v}	−5.72990

of the number of homo- and heteronuclear bonds and other mixing parameters, as a function of composition. Figures 7.11 and 7.12 show the average nearest neighbour distance (ANND), the number of Pd–Pd, Pd–Au and Au–Au bonds and chemical order parameter (σ) as a function of the Au content (m), for the GM found for the average, exp-fit and DFT-fit parameters; at the Gupta level of theory.

The average nearest neighbour distance (ANND) plots (Fig. 7.11) show a steady increase with increasing Au content, due to the larger atomic radius of Au (see Table 7.1). For the average parameters (Fig. 7.11a), there is an increase in the slope at $m = 24$, as for higher Au content the additional Au atoms start to occupy the (111) faces and the core of the cluster—causing a greater rate of the cluster expansion per added Au atom. For the exp-fit parameters (Fig. 7.11b), however, there is a steeper initial slope (as there are more Pd–Au bonds the effect of adding Au atoms is initially greater) but the plot flattens out at $m = 20$. The points at which the slope changes—i.e. $Pd_{14}Au_{24}$ (average) and $Pd_{18}Au_{20}$ (exp-fit)—correspond to the maxima in the Pd–Au bond plot (Fig. 7.12). For the case of the average nearest neighbour distance (ANND) plot for the DFT-fit parameters (Fig. 7.11c), we see a rather jagged behaviour, as this set of parameters leads to high degree of mixing (Pd–Au bonds) but not as much as for the exp-fit parameters.

Figure 7.12a shows that, for the average parameters N_{Pd-Pd} decreases monotonically, while N_{Au-Au} increases, with increasing Au content. N_{Pd-Au} increases to a maximum at $Pd_{14}Au_{24}$, corresponding to the minimum in Δ_{38}^{Gupta} (average) observed in Fig. 7.8. For the exp-fit parameters (Fig. 7.12b), N_{Pd-Pd} decreases monotonically until $m = 20$, after which there are some fluctuations. In contrast to the average parameters, there are no Au–Au bonds up to $Pd_{26}Au_{12}$, as the Au atoms are dispersed on the surface of the cluster (finding a similar behaviour when using the DFT-fit parameters). The number of Pd–Au bonds is maximized at composition $Pd_{18}Au_{20}$, corresponding to the minimum in Δ_{38}^{Gupta} (exp-fit) observed in Fig. 7.8.

Finally, Fig. 7.12 shows plots of the chemical order parameter (σ) for the three sets of parameters. For the average parameters, σ is positive for all compositions, falling to a minimum value of 0 for $Pd_{15}Au_{23}$ and $Pd_{14}Au_{24}$. Positive σ values correspond to core–shell segregation. The fact that the plot goes through a minimum is inevitable because even for core–shell segregation there are a high number of Pd–Au bonds (which lead to a reduction of σ, see Eq. 2.33) for comparable numbers of Pd and Au atoms [12]. The minimum at around $m = 24$ is because of the extra Pd–Au mixing due to the Pd atoms occupying the (111) facets (correlating with the maximum in N_{Pd-Au} shown in Fig. 7.5). For the exp-fit parameters, σ is negative in the region $m = 13-27$ (being 0 for $m = 12$), indicating a high degree of Pd–Au mixing. Similar behaviour is found for the DFT-fit parameters. The minimum in σ for the exp-fit and DFT-fit parameters occurs at $Pd_{18}Au_{20}$, corresponding to the maximum in N_{Pd-Au}, as shown in Fig. 7.12.

It is interesting to note that the minima in the excess energy plots for both the average and fitted parameters correspond to the maximum number of Pd–Au bonds (and minima in σ) for different homotopic configurations. Thus, $Pd_{14}Au_{24}$ (average) has the maximum number of Pd–Au bonds consistent with the $Pd_{core}Au_{shell}$

Fig. 7.11 Variation of average nearest neighbour distance (ANND) as a function of the number of Au atoms (m) for $Pd_{38-m}Au_m$ clusters, for GM obtained using the **a** average, **b** exp-fit and **c** DFT-fit parameters

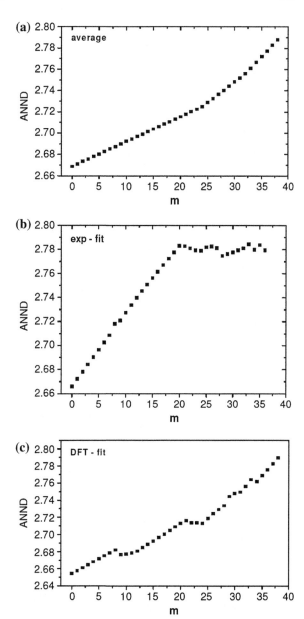

ordering favoured (energetically) by the average potential; while $Pd_{18}Au_{20}$ (fitted) maximises Pd–Au bonding according to the tendency of the fitted parameters to favour more mixed homotops. In principle, the maximum possible number of Pd–Au bonds should be realised for the 1:1 composition, i.e. for $Pd_{19}Au_{19}$, but it is possible that to achieve this maximum Pd–Au bonding would require a structure which is not energetically competitive.

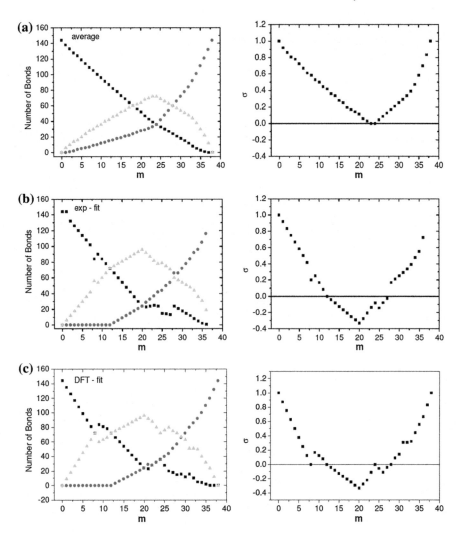

Fig. 7.12 Variation of: (*left*) number of bonds (*black* = Pd–Pd, *red* = Au–Au, *green* = Pd–Au); and (*right*) chemical order parameter (σ), for Pd$_{38-m}$Au$_m$ GM clusters, using the **a** average, **b** exp-fit and **c** DFT-fit parameters

7.6 DFT Investigation of Segregation in 34- and 38-Atom Pd–Au Clusters

Due to the vast combinatorial problem of systematically evaluating all possible homotops for a given nanoalloy structure and composition (see Eq. 1.1) and the impossibility of carrying out a reasonably rigorous homotop search at the DFT level, we have chosen to carry out a DFT analysis of "normal" and "inverted" homotops for the 1:1 composition 34- and 38-atom Pd–Au nanoalloys Pd$_{17}$Au$_{17}$

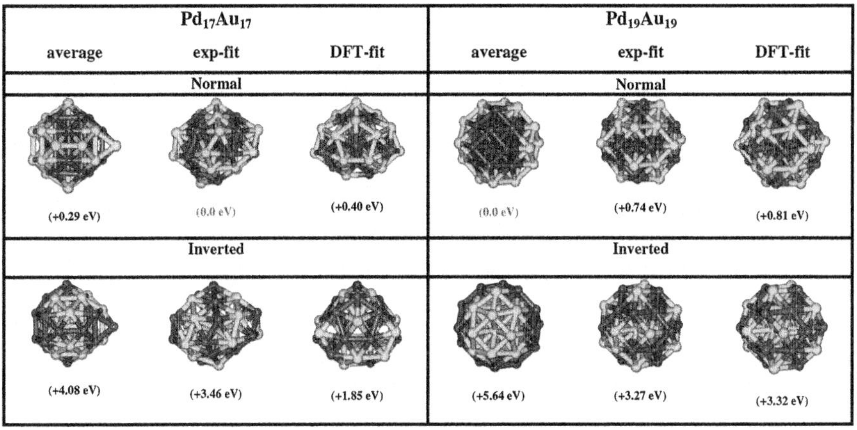

Pd$_{17}$Au$_{17}$			Pd$_{19}$Au$_{19}$		
average	exp-fit	DFT-fit	average	exp-fit	DFT-fit
Normal			Normal		
(+0.29 eV)	(0.0 eV)	(+0.40 eV)	(0.0 eV)	(+0.74 eV)	(+0.81 eV)
Inverted			Inverted		
(+4.08 eV)	(+3.46 eV)	(+1.85 eV)	(+5.64 eV)	(+3.27 eV)	(+3.32 eV)

Fig. 7.13 DFT investigation of segregation effects of 34- and 38-atom Pd–Au clusters, for compositions 17–17 and 19–19, respectively. The "normal" homotop is the lowest energy homotop found using the average or fitted Gupta parameters. In each case, the "inverted" homotop is generated by swapping the positions of all the Pd and Au atoms. The energies quoted are the total DFT cluster energies relative to the lowest energy isomer found for each nuclearity

and Pd$_{19}$Au$_{19}$. This approach has also been used to study a range of nanoalloys clusters [3, 18]. For three sets of Gupta parameters (i.e. average, exp-fit and DFT-fit), the "normal" homotop is the putative GM found (using the BCGA) for Pd$_{17}$Au$_{17}$ or Pd$_{19}$Au$_{19}$. In each case, an "inverted" homotop is generated by exchanging the positions of all Pd and Au atoms. All of the normal and inverted homotops are then energy minimized at the DFT level.

The normal and inverted homotops for Pd$_{17}$Au$_{17}$ and Pd$_{19}$Au$_{19}$ are shown in Fig. 7.13, along with the total DFT cluster energies (in eV) relative to the lowest energy isomer found for each nuclearity. For Pd$_{17}$Au$_{17}$, the isomer derived from the exp-fit parameters (a rather distorted poly-icosahedral pseudo-fivefold symmetry structure, with a high number of Pd–Au bonds) is lower in energy (by 0.29 eV) than the average structure (an incomplete decahedron, with more defined Pd$_{core}$Au$_{shell}$ segregation) after DFT reminimization; which is consistent with its more negative Δ_{34}^{DFT} value (see Fig. 7.6). The DFT-fit parameters predict a similar GM poly-icosahedral structure tp the exp-fit parameters, but this structure is not energetically favourable as it is found to be approximately 0.40 eV higher in energy. DFT optimizations on the three inverted homotops reveal that they are higher in energy (ranging from ~ 1.85 eV for the DFT-fit inverted to ~ 4.08 eV for the average inverted structures). For Pd$_{19}$Au$_{19}$, the "GM" structures for these potentials are truncated octahedra (TO). In this case, the lowest energy structure correspond to that derived from the average parameters (well-defined Pd$_{core}$Au$_{shell}$) which is lower in energy than the exp-fit and DFT-fit parameters predicted structures by 0.74 and 0.81 eV, respectively. This is consistent with the more negative Δ_{38}^{DFT} value (see Fig. 7.10) of the Pd$_{19}$Au$_{19}$ average parameters structure. As the normal and inverted isomers all have TO structures, and the fact that homotop inversion forces Pd and Au atoms to adopt unfavourable sites, it is not surprising that the inverted homotop have higher energies.

It should be noted that, because of the size of the homotop search space, we cannot be sure that we found the lowest energy homotop at the Gupta potential level, let alone at the DFT level. A possible way forward, would be to couple DFT calculations with low temperature Basin Hopping Monte Carlo searching [19], in order to perform a partial homotop optimization by permuting unlike atoms [3, 14–18, 20].

7.7 Large Pd–Au Nanoclusters, $N = 98$ Atoms

7.7.1 Global Optimizations of 98-Atom Pd–Au Clusters

We have studied 98-atom Pd–Au clusters, using a combination of the BCGA and the shell optimization approach (see Chap. 2) using the three different sets of Gupta parameters (i.e. average, exp-fit and DFT-fit). First, we performed a GA search of the potential energy surface (PES) for 98-atom Pd–Au clusters for all compositions $Pd_{98-m}Au_m$. We then calculated the mixing energy of the clusters (Δ_{98}^{Gupta}) for the three sets of parameters. We then focused on the lowest Δ_{98}^{Gupta} composition range ($m = 35-70$). The shell optimization routine was also used to generate all the possible Leary Tetrahedron (LT–T_d) isomers. The objective is to assess how commonly the LT structure is found for 98-atom Pd–Au clusters (as it is the GM structure for 98-atom Pd–Pt clusters, see Chap. 4), as well as to identify other competing low-energy structural motifs.

Using the average parameters, we found that incomplete decahedral structures (labeled Incomplete Dh-Marks) are generally the putative GM over a wide Pd-rich composition range. We also notice a transition to incomplete icosahedral motifs for the Au-rich compositions (i.e. compositions $Pd_{48}Au_{50}$ up to $Pd_{50}Au_{48}$); finding for pure Au_{98} an fcc type structure (see Table 7.8). In terms of segregation, pure Pd_{98} is characterized by an incomplete Dh-Mark structure, and as we increase the concentration of Au in the cluster, Au tends to occupy cluster surface sites; i.e. starts to create "islands" or patches distributed across the cluster surface. The minimum in mixing energy (Δ_{98}^{Gupta}) for the GA searches using the average parameters is found at composition $Pd_{32}Au_{66}$. This is an interesting structure as all the 32 Pd atoms are encapsulated in a distorted (or incomplete) Au_{66} icosahedral cage (see Figs. 7.14 and 7.15).

In contrast, our GA global optimizations using the fitted parameters tend to find clusters in which there is more noticeable Pd–Au mixing (see details about cluster geometries using exp-fit and DFT-fit parameters in Tables 7.9 and 7.10. For the specific case of the exp-fit parameters, we found an fcc type structure (C_1) for pure Pd_{98}. As the Au concentration increases, incomplete icosahedra are found as GM structures (for the Pd-rich composition range). As we start to approach the Au-rich

Table 7.8 Point group symmetries and structures of $Pd_{98-m}Au_m$ GM, using the average parameters, in the $m = 35-70$ composition range

Composition	Symmetry	Structure	Composition	Symmetry	Structure
$Pd_{63}Au_{35}$	C_1	Incomplete Dh-Marks	$Pd_{44}Au_{54}$	C_1	Incomplete Icosahedra
$Pd_{62}Au_{36}$	C_1	Incomplete Dh-Marks	$Pd_{43}Au_{55}$	C_1	Incomplete Dh-Marks
$Pd_{61}Au_{37}$	C_1	Incomplete Dh-Marks	$Pd_{42}Au_{56}$	C_1	Incomplete Dh-Marks
$Pd_{60}Au_{38}$	C_1	Incomplete Dh-Marks	$Pd_{41}Au_{57}$	C_1	Incomplete Dh-Marks
$Pd_{59}Au_{39}$	C_1	Incomplete Dh-Marks	$Pd_{40}Au_{58}$	C_1	Incomplete Dh-Marks
$Pd_{58}Au_{40}$	C_1	Incomplete Dh-Marks	$Pd_{39}Au_{59}$	C_1	Incomplete Dh-Marks
$Pd_{57}Au_{41}$	C_1	Incomplete Dh-Marks	$Pd_{38}Au_{60}$	C_1	Incomplete Icosahedra
$Pd_{56}Au_{42}$	C_1	Incomplete Dh-Marks	$Pd_{37}Au_{61}$	C_1	Incomplete Dh-Marks
$Pd_{55}Au_{43}$	C_1	Incomplete Dh-Marks	$Pd_{36}Au_{62}$	C_1	Incomplete Icosahedra
$Pd_{54}Au_{44}$	C_1	Incomplete Dh-Marks	$Pd_{35}Au_{63}$	C_1	Incomplete Icosahedra
$Pd_{53}Au_{45}$	C_1	Incomplete Dh-Marks	$Pd_{34}Au_{64}$	C_1	Incomplete Icosahedra
$Pd_{52}Au_{46}$	C_1	Incomplete Dh-Marks	$Pd_{33}Au_{65}$	C_1	Incomplete Icosahedra
$Pd_{51}Au_{47}$	C_1	Incomplete Dh-Marks	$Pd_{32}Au_{66}$	C_s	Incomplete Icosahedra
$Pd_{50}Au_{48}$	C_1	Incomplete Dh-Marks	$Pd_{31}Au_{67}$	C_s	Incomplete Icosahedra
$Pd_{49}Au_{49}$	C_1	Incomplete Dh-Marks	$Pd_{30}Au_{68}$	C_s	Incomplete Icosahedra
$Pd_{48}Au_{50}$	C_1	Incomplete Icosahedra	$Pd_{29}Au_{69}$	C_s	Incomplete Icosahedra
$Pd_{47}Au_{51}$	C_1	Incomplete Dh-Marks	$Pd_{28}Au_{70}$	C_1	Incomplete Icosahedra
$Pd_{46}Au_{52}$	C_1	Incomplete Dh-Marks			
$Pd_{45}Au_{53}$	C_1	Incomplete Dh-Marks			

composition range, a structural transition to incomplete-Dh Marks structures takes place at the composition $Pd_{50}Au_{48}$, and continues at least up to composition $Pd_{28}Au_{70}$; a composition still considered to be within the region of lowest Δ_{98}^{Gupta} (see Figs. 7.14 and 7.16). It is interesting to note that both the exp-fit and the DFT-fit parameters predict a larger degree of mixing between Pd and Au atoms compared to the average parameters, due to the larger repulsive (A) and attractive (ζ) terms in the potential (see Fig. 7.14). This is reflected in the lowest mixing energy

average   **exp - fit**   **DFT - fit**

Best GA structures

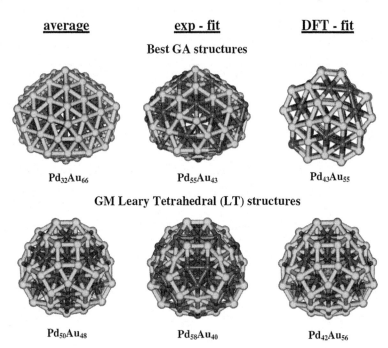

Pd$_{32}$Au$_{66}$ Pd$_{55}$Au$_{43}$ Pd$_{43}$Au$_{55}$

GM Leary Tetrahedral (LT) structures

Pd$_{50}$Au$_{48}$ Pd$_{58}$Au$_{40}$ Pd$_{42}$Au$_{56}$

Fig. 7.14 Lowest $\Delta_{98}^{\text{Gupta}}$ structures found for the three sets of parameters (average, exp-fit and DFT-fit), found during both the BCGA and shell optimization program

$\Delta_{98}^{\text{Gupta}}$ structures for exp-fit (Pd$_{55}$Au$_{43}$) and DFT-fit (Pd$_{43}$Au$_{55}$) structures (see Figs. 7.16 and 7.17), which exhibit marked Pd–Au mixing, though Pd atoms occupy the cores sites.

7.7.2 Leary Tetrahedron Structure

Figure 7.18 shows a dispersion plot of the excess energy ($\Delta_{98}^{\text{Gupta}}$) as a function of Au concentration in the cluster (m) for all the LT (T_d) isomers generated by the shell optimization routine, for the three sets of parameters. From this dispersion plot, we can see that the average parameters give higher $\Delta_{98}^{\text{Gupta}}$ values, and in most cases, most of the isomers lie on the positive side (unfavourable mixing) of $\Delta_{98}^{\text{Gupta}}$, compared to both fitted potentials (exp-fit and DFT-fit). This is due to slightly larger A and ζ values for the fitted potentials, compared to the average potential. For the case of average potential, the great majority of LT (T_d) structures having positive $\Delta_{98}^{\text{Gupta}}$ values are found, to exhibit core–shell type segregation; the unpreferred Au$_{\text{core}}$Pd$_{\text{shell}}$ arrangement (see Chap. 5). For most compositions, LT structures are found to be less stable than the lowest energy structures found during

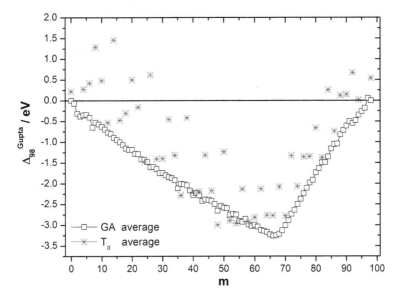

Fig. 7.15 Gupta potential excess energies (Δ_{98}^{Gupta}) for $Pd_{98-m}Au_m$ nanoalloys, as a function of the number of Au atoms (m), over the entire composition range for average parameters. Structures found as putative GM by the GA global optimization are shown in *squares*, while LT T_d structures (lowest energy isomer at each composition) are denoted by *light star shapes*

the GA global optimization. For both the average and exp-fit parameters, the minimum in Δ_{98}^{Gupta} corresponds to a non-LT (T_d) structure, while for the DFT-fit parameters, an LT (T_d) structure (with composition $Pd_{42}Au_{56}$) corresponds to the minimum in Δ_{98}^{Gupta}.

The GA global optimization failed to find LT structures, but the shell optimization program found LT (T_d) structures to be lowest in energy for a small number of compositions: average ($Pd_{74}Au_{24}$, $Pd_{62}Au_{36}$, $Pd_{50}Au_{48}$, $Pd_{46}Au_{52}$ and $Pd_{44}Au_{54}$); exp-fit ($Pd_{58}Au_{40}$ and $Pd_{36}Au_{62}$); and DFT-fit ($Pd_{80}Au_{18}$, $Pd_{46}Au_{52}$, $Pd_{42}Au_{56}$, $Pd_{38}Au_{60}$ and $Pd_{34}Au_{64}$). This indicates that, for Pd–Au clusters, the 98-atom LT structure is less energetically favourable compared to Pd–Pt clusters (see Chap. 4).

7.8 Conclusions

Three parameterisations of the many-body Gupta empirical potential have been compared with regard to the geometrical structures and homotops that they stabilize for Pd–Au nanoalloy clusters with up to 50 atoms, and specific 98-atom size. The "average" parameters (where the Pd–Au parameters are obtained by

Table 7.9 Point group symmetries and structures of $Pd_{98-m}Au_m$ GM, using the exp-fit parameters, in the $m = 35-70$ composition range

Composition	Symmetry	Structure	Composition	Symmetry	Structure
$Pd_{63}Au_{35}$	C_1	Incomplete Icosahedra	$Pd_{44}Au_{54}$	C_1	Incomplete Dh-Marks
$Pd_{62}Au_{36}$	C_1	Incomplete Icosahedra	$Pd_{43}Au_{55}$	C_1	Incomplete Dh-Marks
$Pd_{61}Au_{37}$	C_1	Incomplete Icosahedra	$Pd_{42}Au_{56}$	C_1	Incomplete Dh-Marks
$Pd_{60}Au_{38}$	C_1	Incomplete Icosahedra	$Pd_{41}Au_{57}$	C_1	Incomplete Dh-Marks
$Pd_{59}Au_{39}$	C_1	Incomplete Icosahedra	$Pd_{40}Au_{58}$	C_1	Incomplete Dh-Marks
$Pd_{58}Au_{40}$	C_1	Incomplete Icosahedra	$Pd_{39}Au_{59}$	C_1	Incomplete Dh-Marks
$Pd_{57}Au_{41}$	C_1	Incomplete Icosahedra	$Pd_{38}Au_{60}$	C_1	Incomplete Dh-Marks
$Pd_{56}Au_{42}$	C_1	Incomplete Icosahedra	$Pd_{37}Au_{61}$	C_1	Incomplete Dh-Marks
$Pd_{55}Au_{43}$	C_1	Incomplete Icosahedra	$Pd_{36}Au_{62}$	C_1	Incomplete Dh-Marks
$Pd_{54}Au_{44}$	C_1	Incomplete Icosahedra	$Pd_{35}Au_{63}$	C_1	Incomplete Dh-Marks
$Pd_{53}Au_{45}$	C_1	Incomplete Icosahedra	$Pd_{34}Au_{64}$	C_1	Incomplete Dh-Marks
$Pd_{52}Au_{46}$	C_1	Incomplete Icosahedra	$Pd_{33}Au_{65}$	C_1	Incomplete Dh-Marks
$Pd_{51}Au_{47}$	C_1	Incomplete Icosahedra	$Pd_{32}Au_{66}$	C_1	Incomplete Dh-Marks
$Pd_{50}Au_{48}$	C_1	Incomplete Dh-Marks	$Pd_{31}Au_{67}$	C_1	Incomplete Dh-Marks
$Pd_{49}Au_{49}$	C_1	Incomplete Dh-Marks	$Pd_{30}Au_{68}$	C_1	Incomplete Dh-Marks
$Pd_{48}Au_{50}$	C_1	Incomplete Dh-Marks	$Pd_{29}Au_{69}$	C_1	Incomplete Dh-Marks
$Pd_{47}Au_{51}$	C_1	Incomplete Dh-Marks	$Pd_{28}Au_{70}$	C_1	Incomplete Dh-Marks
$Pd_{46}Au_{52}$	C_1	Incomplete Dh-Marks			
$Pd_{45}Au_{53}$	C_1	Incomplete Dh-Marks			

averaging those for Pd–Pd and Au–Au interactions) are found to favour $Pd_{core}Au_{shell}$ segregation, as these maximize the number of the stronger Pd–Pd bonds and expose the most Au atoms (which have lower surface energies).

The EP preference to segregate Au atoms to surface cluster sites for Pd–Au clusters (i.e. a $Pd_{core}Au_{shell}$ arrangement) can be rationalized as in terms of the following factors: (1) the surface energy of Au is 98 meV/Å whereas that of Pd is 131 meV/Å, (2) the cohesive energies of the two species are roughly the same

Table 7.10 Point group symmetries and structures of $Pd_{98-m}Au_m$ GM, using the DFT-fit parameters, in the $m = 35-70$ composition range

Composition	Symmetry	Structure	Composition	Symmetry	Structure
$Pd_{63}Au_{35}$	C_1	Incomplete Dh-Marks	$Pd_{44}Au_{54}$	C_1	Incomplete Dh-Marks
$Pd_{62}Au_{36}$	C_1	Incomplete Dh-Marks	$Pd_{43}Au_{55}$	C_1	Incomplete Dh-Marks
$Pd_{61}Au_{37}$	C_1	Incomplete Dh-Marks	$Pd_{42}Au_{56}$	C_1	Incomplete Dh-Marks
$Pd_{60}Au_{38}$	C_1	Incomplete Dh-Marks	$Pd_{41}Au_{57}$	C_1	Incomplete Dh-Marks
$Pd_{59}Au_{39}$	C_1	Incomplete Dh-Marks	$Pd_{40}Au_{58}$	C_1	Incomplete Dh-Marks
$Pd_{58}Au_{40}$	C_1	Incomplete Dh-Marks	$Pd_{39}Au_{59}$	C_1	Incomplete Dh-Marks
$Pd_{57}Au_{41}$	C_1	Incomplete Dh-Marks	$Pd_{38}Au_{60}$	C_1	Incomplete Dh-Marks
$Pd_{56}Au_{42}$	C_1	Incomplete Dh-Marks	$Pd_{37}Au_{61}$	C_1	Incomplete Dh-Marks
$Pd_{55}Au_{43}$	C_1	Incomplete Dh-Marks	$Pd_{36}Au_{62}$	C_1	Incomplete Dh-Marks
$Pd_{54}Au_{44}$	C_1	Incomplete Dh-Marks	$Pd_{35}Au_{63}$	C_1	Incomplete Dh-Marks
$Pd_{53}Au_{45}$	C_1	Incomplete Dh-Marks	$Pd_{34}Au_{64}$	C_1	Incomplete Dh-Marks
$Pd_{52}Au_{46}$	C_1	Incomplete Dh-Marks	$Pd_{33}Au_{65}$	C_1	Incomplete Dh-Marks
$Pd_{51}Au_{47}$	C_1	Incomplete Dh-Marks	$Pd_{32}Au_{66}$	C_1	Incomplete Dh-Marks
$Pd_{50}Au_{48}$	C_1	Incomplete Dh-Marks	$Pd_{31}Au_{67}$	C_1	Incomplete Dh-Marks
$Pd_{49}Au_{49}$	C_1	Incomplete Dh-Marks	$Pd_{30}Au_{68}$	C_1	Incomplete Dh-Marks
$Pd_{48}Au_{50}$	C_1	Incomplete Dh-Marks	$Pd_{29}Au_{69}$	C_1	Incomplete Icosahedra
$Pd_{47}Au_{51}$	C_1	Incomplete Dh-Marks	$Pd_{28}Au_{70}$	C_1	Incomplete Icosahedra
$Pd_{46}Au_{52}$	C_1	Incomplete Dh-Marks			
$Pd_{45}Au_{53}$	C_1	Incomplete Dh-Marks			

(3.81 vs. 3.89 eV), respectively, (3) the atomic radius of Au (1.44 Å) is larger than that of Pd (1.38 Å) and finally (4) the Pauling electronegativity of Au (2.4) is slightly larger than that of Pd (2.2), so that any charge transfer from Pd to Au reinforces the tendency to Au surface segregation.

The fitted experimental (exp-fit) and DFT (DFT-fit) parameters (obtained by fitting Pd/Au dissolution energies and the three ordered phases of Pd–Au, respectively) favour Pd–Au mixing, generally leading to mixed Pd–Au nanocluster

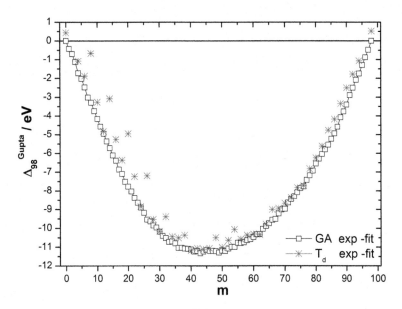

Fig. 7.16 Gupta potential excess energies (Δ_{98}^{Gupta}) for $Pd_{98-m}Au_m$ nanoalloys, as a function of the number of Au atoms (m), over the entire composition range for exp-fit parameters. Structures found as putative GM by the GA global optimization are shown by *squares*, while LT T_d structures (lowest energy isomer at each composition) are denoted by *star shapes*

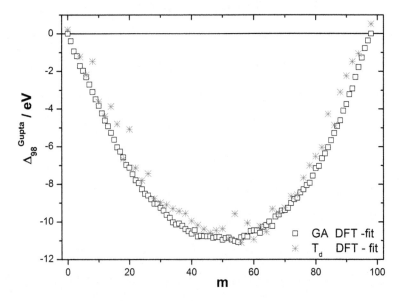

Fig. 7.17 Gupta potential excess energies (Δ_{98}^{Gupta}) for $Pd_{98-m}Au_m$ nanoalloys, as a function of the number of Au atoms (m), over the entire composition range for DFT-fit parameters. Structures found as putative GM by the GA global optimization are shown by *squares*, while LT T_d structures (lowest energy isomer at each composition) are denoted by *star shapes*

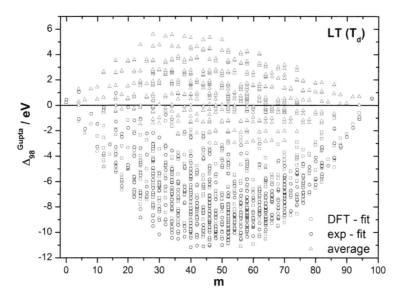

Fig. 7.18 Excess energy (Δ_{98}^{Gupta}) dispersion plot of LT T_d structures, for Pd$_{98-m}$Au$_m$. LT T_d obtained using **a** the average parameters, are denoted by *red triangles*, **b** exp-fit parameters, *black circles* and **c** DFT-fit parameters, *blue squares*

surfaces. Analysis of the 1:1 Pd–Au clusters with up to 50 atoms, shows that there is no clear preference for any one set of parameters over the whole size range. Hence, we decided to compare the three potentials for fixed size (i.e. 34-, 38- and 98-atom size Pd–Au clusters) with variable composition.

From our DFT analysis of 34-atom Pd–Au clusters, excess (mixing) energies (Δ_{34}^{DFT}) curves for the fitted parameters showed a rather jagged behaviour compared with the average parameters. This can be interpreted in the sense that global optimizations (at the EP level) using the fitted parameters tend to find incomplete icosahedra or distorted poly-icosahedra structures, while using average parameters incomplete decahedra motifs are found as the lowest energy structures. It is not clear what the preference is, at the DFT level, for structures based on these potentials; though the exp-fit potential gives marginally lower mixing energies structures, compared to the average and DFT-fit potentials.

For 38-atom nanoalloys, whose structures and chemical ordering have been studied as a function of composition, the average parameters have been found to yield isomers with lower energies at the DFT level, than both fitted parameters (i.e. a preference for more Pd$_{core}$Au$_{shell}$ type structures at this specific size). A more extensive homotop search (incorporating a limited BHMC search and DFT minimization) would be helpful to establish whether the average parameter homotops are in fact the GM or if there are lower energy homotops at the DFT level.

For 98-atom Pd–Au clusters, the GA global optimization searches (using the three sets of parameters) failed to find Leary Tetrahedron (LT) structures as putative GM for any $Pd_{98-m}Au_m$ compositions. LT structures could only be found using the shell optimization program. The complexity of LT potential energy surface (PES) landscape is highlighted by the fact that it was only found to be lower in energy (compared to predicted GA GM structures) at selected compositions; and it was found to be the lowest Δ_{98}^{Gupta} structure only for the DFT-fit parameters.

References

1. R. Ferrando, R.L. Johnston, J. Jellinek, Chem. Rev. **108**, 845 (2008)
2. D.W. Yuan, X.G. Gong, R. Wu, Phys. Rev. B **75**, 085428 (2007)
3. L.O. Paz-Borbón, R.L. Johnston, G. Barcaro, A. Fortunelli, J. Chem. Phys. **128**, 134517 (2008)
4. F. Chen, R.L. Johnston, Acta Mater. **56**, 2374 (2008)
5. D.W. Yuan, X.G. Gong, R. Wu, Phys. Rev. B **78**, 035441 (2008)
6. K. Luo, T. Wei, C.-W. Yi, S. Axnanda, D.W. Goodman, J. Phys. Chem. B **109**, 23517 (2005)
7. S.J. Mejia-Rosales, C. Fernandez-Navarro, E. Pérez-Tijerina, D.A. Blom, L.F. Allard, M. José-Yacamán, J. Phys. Chem. C **111**, 1256 (2007)
8. D. Ferrer, D. Blom, L.F. Allard, S.J. Mejia-Rosales, E. Pérez-Tijerina, M. José-Yacamán, J. Mater. Chem. **18**, 2442 (2008)
9. M. José-Yacamán, S.J. Mejia-Rosales, E. Pérez-Tijerina, J. Mater. Chem. **17**, 1035 (2007)
10. E. Pérez-Tijerina, M. Gracia-Pinilla, S.J. Mejia-Rosales, U. Ortiz-Méndez, A. Torres, M. José-Yacamán, Faraday Discuss. **138**, 353 (2008)
11. C. Massen, T.V. Mortimer-Jones, R.L. Johnston, J. Chem. Soc. Dalton Trans. **23**, 4375 (2002)
12. L.O. Paz-Borbón, A. Gupta, R.L. Johnston, J. Mater. Chem. **18**, 4154 (2008)
13. R. Hultgren, P.D. Desai, D.T. Hawkins, M. Gleiser, K.K. Kelley, *Values of the Thermodynamic Properties of Binary Alloys* (American Society for Metals, Jossey-Bass Publishers, Berkeley, 1981)
14. L.O. Paz-Borbón, T.V. Mortimer-Jones, R.L. Johnston, A. Posada-Amarillas, G. Barcaro, A. Fortunelli, Phys. Chem. Chem. Phys. **9**, 5202 (2007)
15. G. Rossi, R. Ferrando, A. Rapallo, A. Fortunelli, B.C. Curley, L.D. Lloyd, R.L. Johnston, J. Phys. Chem. **122**, 194309 (2005)
16. A. Rapallo, G. Rossi, R. Ferrando, A. Fortunelli, B.C. Curley, L.D. Lloyd, G.M Tarbuck, R.L. Johnston, J. Chem. Phys. **122**, 194308 (2005)
17. B.C. Curley, R.L. Johnston, G. Rossi, R. Ferrando, Eur. Phys. J. D. **43**, 53 (2007)
18. L.O. Paz-Borbón, R.L. Johnston, G. Barcaro, A. Fortunelli, J. Phys. Chem. C **111**, 2936 (2007)
19. J.P.K. Doye, D.J. Wales, J. Phys. Chem. A **101**, 5111 (1997)
20. G. Rossi, A. Rapallo, C Mottet, A Fortunelli, F. Baletto, R. Ferrando, Phys. Rev. Lett. **93**, 105503 (2004)

Chapter 8
Chemisorption on Metal Clusters and Nanoalloys

8.1 Introduction

One of the important characteristics of nanoparticles is their high surface/volume ratio. This makes them very attractive as catalysts, as most chemical reactions occur at surface sites. An important difference between studying reactions on a finite cluster (both experimentally and computationally), rather than on a periodic infinite surface, is that clusters often present chemically active coordination sites which are not present on an ideal extended surface. The nature and number of active surface sites also vary with particle shape, packing and size.

8.1.1 Experimental and Theoretical Studies on Chemisorption

There have been many experimental studies of CO adsorption on metallic surfaces using various techniques, such as EELS, thermal desorption spectroscopy, EXAFS, IR spectroscopy, X-Ray diffraction, dynamical LEED intensity analysis and STM [1–6]. These results agree in that a-top positions, along with bridge bonds are the best adsorption sites for CO molecules on Pt surfaces, while for the case of Pd surfaces a preference is found for bridging positions [7–9].

Feibelman et al. performed extensive DFT calculations highlighting the importance of how the different approximations used in DFT theory (e.g. pseudo-potentials, exchange-correlation functional, basis set size) can lead to results in which different adsorption sites are preferred (i.e. it was called the CO/Pt(111) "puzzle") [10]. Other metals, such as Au (which is know to be characterized by a endothermic chemisorption energy) have been found to present unique catalytic properties in nanoparticles of a few nanometers diameter, as found by Hvolbaek, et al. [11]. They used an Au_{10} cluster model in order to model the CO oxidation process. Two alternative pathways were studied: (a) O_2 dissociates before reacting

L. O. Paz Borbón, *Computational Studies of Transition Metal Nanoalloys*, Springer Theses, DOI: 10.1007/978-3-642-18012-5_8, © Springer-Verlag Berlin Heidelberg 2011

with CO to form CO_2; (b) direct reaction between the O_2 and CO. Their results showed that low-coordinated Au atoms are able to bind the CO and O (a prerequisite for a catalytic reaction) so the reaction between O_2 and CO is favourable; as it requires a lower activation energy than the reaction involving O_2 dissociation.

DFT calculations on H adsorption on small Pt clusters have shown that dissociative H_2 chemisorption energies and H desorption are strongly coverage dependent, showing a general decrease with increasing hydrogen coverage, with the a-top position preferred over edge or hollow sites [12]. Chen et al. performed DFT calculations on a Pt_6 cluster and its interactions with H_2, finding a preference for a-top positions, sequential dissociative chemisorption of H_2 and that this adsorption is governed by charge transfer effects from the Pt cluster to the hydrogen molecules ($5d/1\sigma^*$ orbital overlap) [13]. Calvo and Carré have studied the stability of Pd icosahedral and cuboctahedral clusters under H adsorption by means of Monte Carlo simulations using empirical potentials. Hydrogen concentrations close to 75% coverage are found to favour cuboctahedral structures relative to icosahedral ones, suggesting a structural transition induced by H adsorption [14]. Neyman and coworkers [15, 16] studied the adsorption site-energetics of CO on Pd 1–2 nm particles via DFT calculations, finding results in good agreement with experimental data in UHV. In recent work [17], morphology changes in Au clusters induced by CO adsorption have been predicted. Thus, high-level computational techniques, such as DFT, have been shown to play an important role in understanding the effect of atomic and electronic structure on the chemisorption of small molecules on nanoparticles.

We have performed a theoretical study of pure nanoparticles (Pd/Pt/Au) and their interaction with CO molecules and atomic H, in order to study chemisorption effects of relevance to nanoparticle catalysis. First-principles density functional (DFT) local-relaxations are used to investigate the effect of CO and H adsorption on six structural motifs. The results of the energetic crossover and structural deformations are analyzed in terms of the interplay between metal–metal interactions (including internal and surface stress) and CO–metal and H–metal interactions.

8.2 Chemisorption of CO and H on Pd, Pt and Au Nanoclusters

Six structural motifs for Pd, Pt and Au clusters have been found from a previous thorough investigation [18] to be the most competitive in terms of energy, and used in this study. These structural motifs, which are shown in Fig. 8.1, can be described as follows: (a) Ih-Mc_{38} (38 atoms) is a fragment of a 55-atom Mackay icosahedron; (b) Dh_{39} (39 atoms) is a fragment of a 55-atom decahedron; (c) aMc (40 and 39 atoms) is a polyicosahedral (anti-Mackay) motif with fivefold symmetry. aMc_{40} is the original structure [19] while aMc_{39} is a distorted decahedral type configuration resulting from the local optimization of the aMc_{40} Au cluster

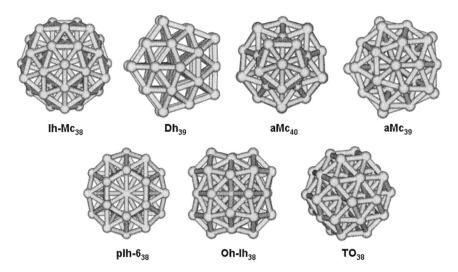

Fig. 8.1 Structural families (motifs) used as model Pd, Pt and Au clusters

which has a single atom protruding out of the cluster tip: the aMc$_{39}$ structure is obtained by removing the segregated atom.; (d) TO$_{38}$ (38 aTOms) is a truncated octahedron; (e) pIh − 6$_{38}$ (38 atoms) is a polyicosahedral motif with sixfold symmetry [20]; (f) Oh-Ih$_{38}$ (38 atoms) is a hybrid fivefold-symmetry/close-packed structure [18]. In order to speed up calculations, only motifs with geometric shell-closure have been considered.

8.2.1 Energetic Analysis

The structures shown in Fig. 8.1 were fully covered with H or CO, adsorbed in a-top positions on all the surface metal atoms, and subjected to local energy minimizations at the DFT level.

The *relative total energy* (ΔE_{Total}) at the DFT level, for a bare or chemisorbed cluster is calculated as:

$$\Delta E_{Total} = E_{Total}(M_{38}) - E_{Total}(M_{38}, GM) \tag{8.1}$$

where $E_{Total}(M_{38}, GM)$ is the total DFT energy of the lowest energy bare metal 38-atom cluster. As some of the structures considered have different numbers of metal atoms, to enable a comparison of their energies these energies must be rescaled. For sizes $N = 39$ and $N = 40$, $E_{Total}(M_N)$ must be rescaled to give $E^*_{Total}(M_N)$ $= \frac{E_{Total}(M_N) \cdot 38}{N}$. For the chemisorbed clusters ($M_N X_{32}$, where X_{32} refers to 32 H atoms or 32 CO molecules, at sizes $N = 39$, 40; the energy is rescaled as follows:

$$E^*_{Total}(M_N X_{32}) = E_{Total}(M_N X_{32}) - E_{Total}(M_N) + E^*_{Total}(M_N) \tag{8.2}$$

These differences in total DFT energies (ΔE_{Total}) are plotted in Figs. 8.2, 8.3 and 8.4 for Pd, Pt and Au.

The *interaction energy* per adsorbate atom or molecule, E_{int}, for the chemisorption of 32 H atoms or 32 CO molecules on $N = 38$, 39 or 40 metal clusters are calculated using the following expression:

$$E_{int}(M_N X_{32}) = \frac{1}{32}[E_{Total}(M_N X_{32}) - E_{Total}(M_N) - 32 \cdot (E_{Total}(X))] \qquad (8.3)$$

while selective adsorption (i.e. variable numbers, X_m of H atoms or CO molecules), is defined as:

$$E_{int}(M_N X_m) = \frac{1}{m}[E_{Total}(M_N X_m) - E_{Total}(M_N) - m \cdot (E_{Total}(X))] \qquad (8.4)$$

Interaction energies are defined as negative quantities. Finally, the cluster *distortion energy*, E_{dist} is defined as the energy difference between the structurally relaxed bare metal cluster [with energy $E_{Total}(M_N)$] and the energy $[E_{Total}^{frozen}(M_N : M_N X_{32})]$ of the metal cluster core of the chemisorbed cluster, with the adsorbates removed but the core geometry frozen:

$$E_{dist} = E_{Total}^{frozen}(M_N : M_N X_{32}) - E_{Total}(M_N) \qquad (8.5)$$

Distortion energies are positive by definition, as the frozen metal core of a $M_N X_{32}$ cluster lies at higher energy than the relaxed cluster.

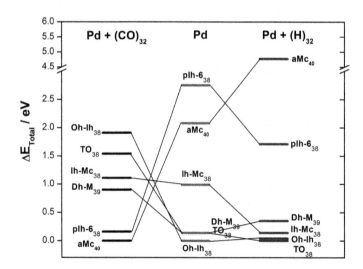

Fig. 8.2 Comparison of relative energy $\Delta E_{Total}(M_{38})$ of structural motifs for Pd clusters before and after a-top chemisorption of 32 CO molecules or H atoms

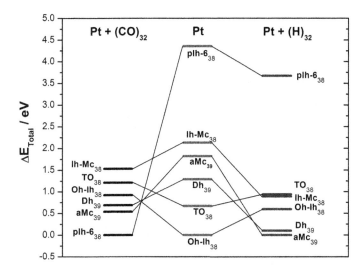

Fig. 8.3 Comparison of relative energies $\Delta E_{Total}(M_{38})$ of structural motifs for Pt clusters before and after a-top chemisorption of 32 CO molecules or H atoms. $E_{Total}(M_{38})$ is rescaled for 39- and 40-atom clusters

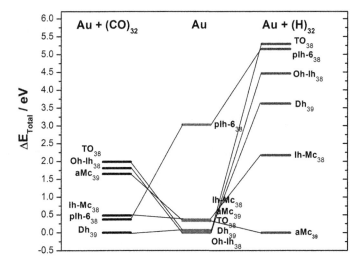

Fig. 8.4 Comparison of relative energy $\Delta E_{Total}(M_{38})$ of structural motifs for Au clusters before and after a-top chemisorption of 32 CO molecules or H atoms

8.2.2   Computational Details

DFT calculations were performed using the NWChem 5.0 quantum chemistry software package [21] and the PW91 exchange-correlation functional [22]. Computational details are identical to those described in Chap. 2, except for the basis sets for the light atoms, chosen as standard $H(4s)/[2s]$; C and O $(8s4p)/[4s2p]$ basis sets for the structural optimizations [23]. After the optimization step, all the energy values are reported using a triple-zeta-plus-polarization basis set, as in described in Chap. 2 for metal atoms; with the addition of $H(5s1p)/[3s1p]$ and C and O $(11s6p1d)/[5s3p1d]$ basis sets [23]. All calculations were performed spin unrestricted.

8.3 Results

The DFT energy ordering of the structural motifs, before and after chemisorption of 32 CO molecules or H atoms (i.e. full coverage of a-top sites) are compared in Fig. 8.2 for Pd, Fig. 8.3 for Pt and Fig. 8.4 for Au. Tables 8.1 and 8.2 show the corresponding interaction and distortion energies for these three metal systems. One can immediately observe very different qualitative behaviour upon H and CO adsorption, so these are analyzed separately below.

For bare Au clusters, all the structural motifs are rather close in energy, except for $pIh - 6_{38}$, which suffers from internal strain [24]. Upon H adsorption, the spread of energies is much increased, with the energy differences between chemisorbed and bare clusters being in the order: $aMc_{39} \gg pIh - 6_{38} \approx Ih\text{-}Mc_{38} > Dh_{39} > Oh\text{-}Ih_{38} > TO_{38}$. Remembering that fcc Au does not interact favourably with hydrogen (gold is the only metal that does not dissociate the H_2 molecule in macroscopic form), the effect of H adsorption can then be easily rationalized in terms of larger interactions for the more exposed (less coordinated or less "fcc-like") sites.

On the other hand, Pt, being a very "sticky" metal, tends to disfavor structures entailing substantial surface stress, such as $Ih\text{-}Mc_{38}$ or Dh_{39}. With respect to Au, it interacts more strongly with hydrogen, as shown in Table 8.1 (the interaction energies tend to be similar for different motifs). The effect of H adsorption is to increase coordination of surface atoms and to decrease their stress, thus favouring $Ih\text{-}Mc_{38}$ and Dh_{39} with respect to less strained configurations. This is consistent with the initial favouring of icosahedra with respect to fcc clusters at low H content (occupation of outer sites) in Ref. [14]. $pIh - 6_{38}$ is a different case,

Table 8.1 Interaction energies (E_{int}/eV) for H/CO adsorption

Motif	Pd	Pt	Au
$Oh\text{-}Ih_{38}$	−2.19/−1.42	−2.67/−1.84	−1.97/−0.61
TO_{38}	−2.22/−1.46	−2.68/−1.85	−1.95/−0.61
Dh_{39}	−2.19/−1.45	−2.73/−1.89	−2.00/−0.67
$Ih - M_{38}$	−2.22/−1.47	−2.73/−1.89	−2.01/−0.67
$pIh - 6_{38}$	−2.21/−1.54	−2.71/−2.01	−2.04/−0.75
aMc_{40}	−2.13/−1.56	−/−	−/−
aMc_{39}	−/−	−2.75/−1.91	−2.12/−0.63

Dashes indicate DFT–SCF failure, probably due to selected spin state configuration

Table 8.2 Distortion energies (eV) for H/CO adsorption

Motif	Pd	Pt	Au
Oh-Ih$_{38}$	0.63/1.36	1.52/3.58	2.69/1.78
TO$_{38}$	0.41/1.42	1.04/4.03	1.97/2.09
Dh$_{39}$	0.48/1.34	0.99/3.01	1.08/1.56
Ih − M$_{38}$	0.57/1.29	1.26/3.24	2.41/1.51
pIh − 6$_{38}$	0.16/1.15	1.81/3.90	1.16/2.07
aMc$_{40}$	0.32/1.20	−/−	−/−
aMc$_{39}$	−/−	2.01/4.00	3.58/1.25

Dashes indicate DFT-SCF failure, probably due to selected spin state configuration

because H coordination cannot reduce its internal stress. Finally, Pd is similar to Pt, but with the difference that, as Pd is a less "sticky" metal, the TO$_{38}$ structure does not correspond to an ideal shape for Pd, so that it is also stabilized significantly by H-coverage.

In contrast to hydrogen, the interaction with CO molecules appreciably changes the cluster potential energy landscape, with non-crystalline structures being stabilized relative to fcc-type motifs. This can be better appreciated by an analysis of the distortion energies (E_{dist}) listed in Table 8.2. The main effect that one can observe from this table is a much larger (more than doubled) value of the distortion energies for CO with respect to H for Pd and Pt clusters. This is because CO interacts much more strongly with Pd and Pt atoms via the σ-donation/π-backdonation mechanism [17, 15]. Substantial CO coordination, such as considered here (see Fig. 8.5), thus brings about a partial disruption of the metal–metal bonds, and a "flattening" of the potential energy surface of the metal clusters, with energy differences between them reduced to within 2 eV. The behaviour of Au is

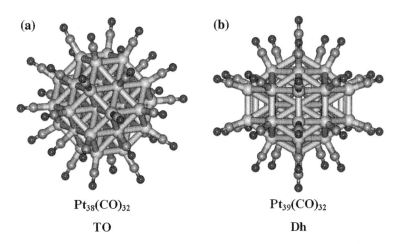

(a)	(b)
Pt$_{38}$(CO)$_{32}$	Pt$_{39}$(CO)$_{32}$
TO	Dh

Fig. 8.5 Structures of **a** TO$_{38}$ (Pt$_{38}$) and **b** Dh$_{39}$ (Pt$_{39}$) with 32 CO molecules adsorbed on surface a-top sites

somewhat different, in that fcc-like structures (TO_{38} and $Oh-Ih_{38}$) interact less strongly with CO because of increased coordination and are thus destabilized.

A more detailed analysis confirms the previous description. The interaction energies per adsorbate unit of TO_{38}, $pIh - 6_{38}$ and Dh_{39} Pt clusters are plotted against the number of a-top adsorbate ligands in Fig. 8.6 for CO (only TO_{38} and $pIh - 6_{38}$ structures) and Fig. 8.7 for H-adsorption; while Fig. 8.8 shows the interaction energies of TO_{38} (for selective adsorption of CO and H) for Pd, Pt and Au clusters.

The choice of adsorption sites (always full occupation of sites equivalent in symmetry) is the following: for TO_{38}, 8 (111)-centroid or 24 vertex sites; for Dh_{39},

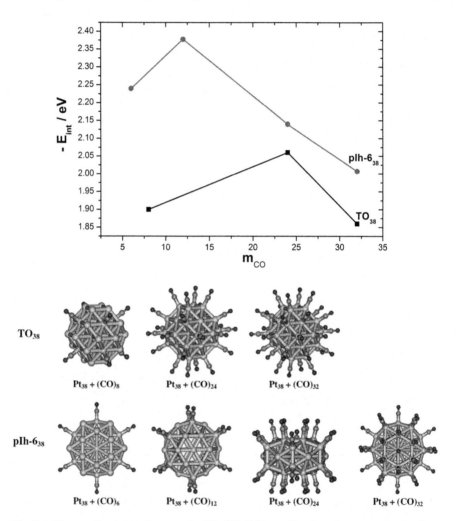

Fig. 8.6 *Top* negative interaction energies $[E_{int}(M_N X_m)]$ as a function of coverage, m, for three structural motifs of Pt_{38} with CO. *Bottom* Schematic representation of $pIh - 6_{38}$ and TO_{38} structures, as a function of CO coverage. Figure *(top)* only shows two structural motifs due to SCF convergence failure of Dh_{39} structure

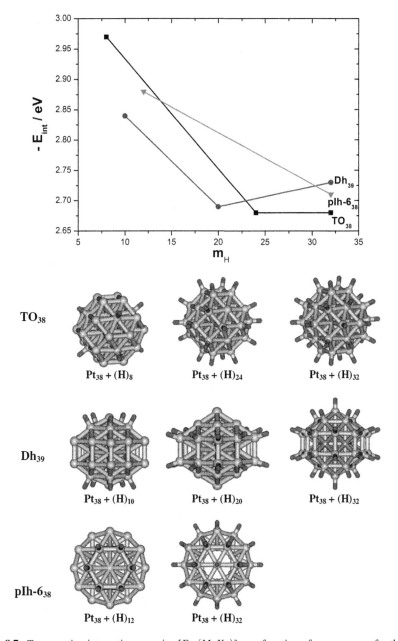

Fig. 8.7 *Top* negative interaction energies $[E_{int}(M_N X_m)]$ as a function of coverage, m, for three structural motifs of Pt$_{38}$ with H. *Bottom* schematic representation of pIh $- 6_{38}$, Dh$_{39}$ and TO$_{38}$ structures, as a function of H coverage

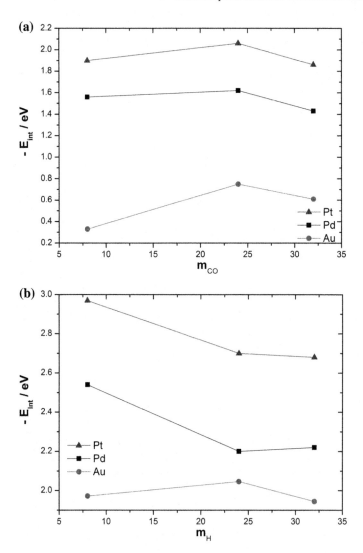

Fig. 8.8 Negative interaction energies $[E_{int}(M_N X_m)]$ as a function of coverage, m, for the 38-atom TO_{38} motif for Pd, Pt and Au clusters with **a** CO and **b** H

10 sites on edges connecting apexes to off-axis vertexes or 20 off-axis vertex sites; for pIh -6_{38}, 6 sites in the equatorial symmetry plane, 12 sites off the symmetry plane (but close to it) or 24 sites consisting of the previous 12 + 12 sites further from the equatorial plane.

In the case of H-adsorption, comparing the decahedral (Dh_{39}) structure with the truncated octahedral (TO_{38}) motif, we notice that the distortion energies are similar, see Table 8.2. Figure 8.7 shows that the last 12 H atoms (i.e. going from $Pt_{38}H_{20}$ to $Pt_{38}H_{32}$) allow the structures to relax. The TO_{38} structure undergoes a

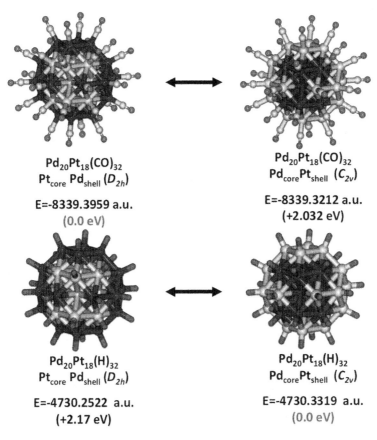

Pd$_{20}$Pt$_{18}$(CO)$_{32}$
Pt$_{core}$ Pd$_{shell}$ (D_{2h})

E=-8339.3959 a.u.
(0.0 eV)

Pd$_{20}$Pt$_{18}$(CO)$_{32}$
Pd$_{core}$Pt$_{shell}$ (C_{2v})

E=-8339.3212 a.u.
(+2.032 eV)

Pd$_{20}$Pt$_{18}$(H)$_{32}$
Pt$_{core}$ Pd$_{shell}$ (D_{2h})

E=-4730.2522 a.u.
(+2.17 eV)

Pd$_{20}$Pt$_{18}$(H)$_{32}$
Pd$_{core}$Pt$_{shell}$ (C_{2v})

E=-4730.3319 a.u.
(0.0 eV)

Fig. 8.9 Total and relative DFT energies for Pt$_{core}$Pd$_{shell}$ and Pd$_{core}$Pt$_{shell}$ bimetallic Pd$_{20}$Pt$_{18}$TO$_{38}$ cluster, with 32 adsorbed H atoms or CO molecules

small relaxation of the (111) facets, while the Dh$_{39}$ structure undergoes a larger reconstruction, leading to a new more rounded shape. It is known that the Dh$_{39}$ structure has internal elastic strains, surface Pt coordination, allows this strain to be released, via expansion of the radial Pt–Pt bonds. Similar results have also been found for Pd$_{38}$H$_m$ clusters. Turning now to CO-adsorption, we observe in Fig. 8.6 a smoother behaviour of the ligand interaction energies as a function of coverage, consistent with previous results [17, 15]. Moreover, from the structural point of view a significant expansion of the central octahedral core of the TO$_{38}$ upon CO adsorption is found, with the Pt–Pt bond length increasing from 2.78 Å to 2.92 Å, and disappearance of the protusion of atoms on (111) facets, a sign of the decreased importance of directionality effects [24], see Fig. 8.5. For the polyico-sahedral (pIh − 6$_{38}$) structure, CO-adsorption induces a compression along the sixfold symmetry axis, and an expansion in the plane perpendicular to this axis, thereby increasing the eccentricity or oblateness of the cluster. Again, there is an overall inflation of the cluster and a weakening of the metal–metal bonds.

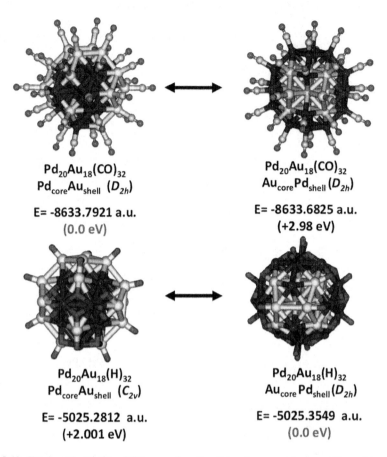

Fig. 8.10 Total and relative DFT energies for $Pd_{core}Au_{shell}$ and $Au_{core}Pd_{shell}$ bimetallic $Pd_{20}Au_{18}TO_{38}$ cluster, with 32 adsorbed H atoms or CO molecules

8.4 Chemisorption of CO and H on Bimetallic Clusters

Bimetallic systems (solids, surfaces and nanoparticles) usually present modified chemical and physical properties compared to the pure metals. One can distinguish two different types of bimetallic systems: one in which the chemical and physical properties lie outside those of the single components; and the other where the properties lie in between. These effects are usually explained in terms of a direct electronic interaction between the two metals, along with geometric effects (i.e. different lattice constants) [25].

According to recent DFT calculations, if a more reactive metal with a smaller lattice constant is mixed with a more inert metal having a larger lattice constant (such as Ru and Pt or Pd and Au) then a bimetallic system which binds adsorbates more strongly than either pure componenets will result. On the other hand, if a more reactive metal with a large lattice constant is mixed with a more inert metal with small lattice constant (e.g. Pd and Cu) then, an almost homogeneous metal with intermediate properties will be obtained [26].

As an example, it has been found that Pd–Au nanoparticles can have catalytic properties distinct from those of pure Pd and Au particles, due to their differing electronic structures: e.g. they are catalysts for the low temperature synthesis of hydrogen peroxide and for CO and alcohol reduction [27, 28]. Finally, in a recent experimental study of H_2 adsorption on bimetallic $Pd_{core}Pt_{shell}$ nanoparticles, Kobayashi et al. found that, after dissociation, the H atoms penetrate into the Pd–Pt particles and are trapped at the Pd–Pt interface. This novel result highlights the importance of theoretical studies of bimetallic nanoparticles not only for they catalytic properties, but also for possible hydrogen storage applications [29].

While the global minimum (GM) structure for 38-atom bimetallic clusters is found to depend on composition, as well as on the particular bimetallic system, (e.g. Pd–Au or Pd–Pt) we have found (see Sect. 8.5) that surface segregation is mainly determined by the minimization of surface energy and the maximization of the strongest bond interactions: $Pt_{core}Pd_{shell}$ and $Pd_{core}Au_{shell}$ are the preferred configurations for these two systems.

On the other hand, the previously predicted $Pt_{core}Pd_{shell}$ segregation is modified by the presence of H atoms (full cluster coverage), for which the $Pd_{core}Pt_{shell}$ configuration is preferred, whereas CO adsorption (full cluster coverage) does not have a significant effect on Pd surface segregation.

For Pd–Au clusters, H adsorption also modifies the predicted bare cluster segregation (i.e. $Pd_{core}Au_{shell}$) with $Au_{core}Pd_{shell}$ segregation now being preferred. It is also noticeable that H adsorption induces a structural re-arrangement of the cluster, as the initial fcc type (TO_{38}) motif is distorted to an asymmetrical structure, while the H atoms now occupy some of the (111) hollow sites between the Au and Pd atoms. These effects are not well understood, but mixing of Pd and Au, as well as Pt and Pd (i.e. alloying effects as a function of concentration of one particular metal in the nanoparticle) plays an important role when determining the catalytic activity and properties of these nanoalloys. It is interesting to note that adsorption induced inversion of core–shell ordering has been observed experimentally by e.g. Kiely et al. (for Pd–Au nanoparticles [28]) and by Somorjai and colleagues (for Rh–Pt nanoparticles [30]) (Figs. 8.9, 8.10).

8.5 Conclusions

The DFT energy ordering of the structural motifs of Pd, Pt and Au clusters of size 38, has been found to be altered by CO and H chemisorption. CO chemisorption has been found to induce a "flattening" of the metal cluster PES, for the three systems considered in this study, favoring non-crystalline (e.g., polyicosahedral) structures, whereas the effect of H adsorption is mostly connected with the release of surface stress. These findings are in tune with experimental observations of structural changes in catalytic conditions [31–33]. Moreover, these general trends could be further supported by an electronic structure analysis, quantifying, e.g. charge transfer between the metal clusters and the adsorbed molecules or the changes in the local density of states of the metal d-shells [17, 15].

For the case of bimetallic clusters, we performed DFT calculations for full coverage of H atoms and CO molecules (32 in total) at one selective composition: (a) $Pd_{20}Au_{18}$ for both $Pd_{core}Au_{shell}$ and $Au_{core}Pd_{shell}$ configurations; and (b) $Pd_{20}Pt_{18}$ for *both*$Pt_{core}Pd_{shell}$ and $Pd_{core}Pt_{shell}$. It is interesting to note that the adsorption of CO did not affect the predicted segregation for these particles (i.e. $Pd_{core}Au_{shell}$ and $Pt_{core}Pd_{shell}$), nor their structures. Adsorption of H, however, strongly affected the Pd–Au cluster structure and the final adsorption sites of H atoms on the nanoparticle surface. H adsorption also affected the segregation properties for both Pd–Au and Pd–Pt clusters (preferring the $Au_{core}Pd_{shell}$ and $Pd_{core}Pt_{shell}$ configurations), which can be explained in terms of higher Pt–H and Pd–H interaction energies compared to Au–H. More first-principles calculations are needed (e.g. exploring different adsorption sites such as "bridge" or (111) hollow sites, as well as selective adsorption) for bimetallic systems in order to get a full understanding of catalytic properties as a function of nanoparticle composition and chemical ordering.

References

1. H. Hopster, H. Ibach (1978) Surf. Sci. **77**, 109
2. S.L. Anderson, T. Mizushima, Y. Udagawa (1991) J. Phys. Chem. **95**, 6603
3. B.E. Hayden, K. Kretzschmar, A.M. Bradshaw, Surf. Sci. **149**, 394 (1985)
4. D.F. Ogletree, M.A. Van Hove, G.A. Somorjai, Surf Sci. **173**, 351 (1986)
5. G.S. Blackman, M.-L. Xu, D.F. Ogletree, M.A. Van Hove, G.A. Somorjai, Phys. Rev. Lett. **61**, 2352 (1988)
6. M. Pedersen, M.-L. Bocquet, P. Sautet, E. Lgsgaard, I. Stensgaard, F. Besenbacher (1999) Chem. Phys. Lett. **299**, 403
7. R.J. Behm, K. Christmann, G. Ertl , M.A. Van Hove (1980) J. Chem. Phys. **73**, 2984
8. H. Conrad, G. Ertl, J. Koch, E.E. Latta, Surf. Sci. **43**, 462 (1974)
9. A.M. Bradshaw, F.M. Hoffmann, Surf. Sci. **72**, 513 (1978)
10. P.J. Feibelman, B. Hammer, J.K. Nørskov, F. Wagner, M. Scheffler, R. Stumpf, R. Watwe, J. Dumesic, J. Phys. Chem. B **105**, 4018 (2001)
11. B. Hvolbæk, T.V.W. Janssens, B.S. Clausen, H. Falsig, C.H. Christensen, J.K. Nørskov (2007) Nanotoday **2**, 4–14
12. C. Zhou, J. Wu, A. Nie, R.C. Forrey, A. Tachibana, H. Cheng, J. Phys. Chem. C **111**, 12773 (2007)
13. L. Chen, A.C. Cooper, G.P. Pez, H. Cheng, J. Phys. Chem. C **111**, 5514 (2007)
14. F. Calvo, A. Carré, Nanotechnology **17**, 1292 (2006)
15. I.V. Yudanov, R. Sahnoun, K.M. Neyman, N. Rösch J. Chem. Phys. **117**, 9887 (2002)
16. I.V. Yudanov, R. Sahnoun, K.M. Neyman, N. Rolsch, J. Hoffmann, S. Schauermann, V. Johánek, H. Unterhalt, G. Rupprechter, J. Libuda, H.-J. Freund J. Phys. Chem. B **107**, 255 (2003)
17. K. McKenna, A.L. Schluger, J. Phys. Chem. C **111**, 18848 (2007)
18. L.O. Paz-Borbón, R.L. Johnston, G. Barcaro, A. Fortunelli, J. Chem. Phys. **128**, 134517 (2008)

19. G. Barcaro, A. Fortunelli, F. Nita, G. Rossi, R. Ferrando, J. Phys. Chem. B **110**, 23197 (2006)
20. G. Rossi, A. Rapallo, C. Mottet, A. Fortunelli, F. Baletto, R. Ferrando, Phys. Rev. Lett. **93**, 105503 (2004)
21. R.A. Kendall, E. Aprà, D.E. Bernholdt, E.J. Bylaska, M. Dupuis, G.I. Fann, R.J. Harrison, J. Ju, Nichols, J. Nieplocha, T.P. Straatsma, T.L. Windus, A.T. Wong, Comput. Phys. Commun. **128**, 260 (2000)
22. J.P. Perdew, J.A. Chevary, S.H. Vosko, K.A. Jackson, M.R. Pederson, D.J. Singh, C. Fiolhaus, Phys. Rev. B **46**, 6671 (1992)
23. See: ftp://ftp.chemie.uni-karlsruhe.de/pub/basen/
24. R. Ferrando, A. Fortunelli, G. Rossi, Phys. Rev. B **72**, 085449 (2005)
25. R. Ferrando, R.L. Johnston, J. Jellinek, Chem. Rev. **108**, 845 (2008)
26. A. Gross, Topic. Catal. **37**, 29 (2006)
27. J.K. Edwards, B.E. Solsona, P. Landon, A.F. Carley, A. Herzing, C.J. Kiely, G.J. Hutchings, J. Catal. **236**, 69 (2005)
28. A.A. Herzing, M. Watanabe, J.K. Edwards, M. Conte, Z.-R.Tang, G.J. Hutchings, C.J. Kiely, Faraday Discuss. **138**, 337 (2008)
29. H. Kobayashi, M. Yamauchi, H. Kitagawa, Y. Kubota, K. Kato, M. Takata, J. Am. Chem. Soc. **130**, 1818 (2008)
30. F. Tao, M.E. Grass, Y.W. Zhang, D.R. Butcher, J.R. Renzas, Z. Liu, J.Y. Chung, B.S. Mun, M. Salmeron, G.A.Somorjai, Science **322**, 932 (2008)
31. M.A. Newton, C. Belver-Coldeira, A. Martnez-Arias, M. Fernndez-Garca, Nature Mater. **6**, 528 (2007)
32. J. Evans, J. Tromp, J. Phys.: Cond. Mat. **20**, 184020 (2008)
33. G. Rupprechter, C. Weilhach, J. Phys.: Cond. Mat. **20**, 184019 (2008)

Chapter 9
Conclusions and Future Work

In this thesis we have shown that the combination of empirical potentials (EP), such as the Gupta potential, with DFT calculations, and employing a generic algorithm (GA) and Basin Hopping Monte Carlo (BHMC) techniques, is a powerful tool for identifying low energy structural motifs and segregation patterns in bimetallic alloys.

By combining Gupta potential global optimizations and DFT local energy minimisations, we have identified a new particularly stable structural motif for 34-atom Pd–Pt clusters, the Dh-cp(DT) structure. Moreover, the energetic ordering of global minimum structural motifs predicted by the Gupta potential is not confirmed at the higher level of theory (i.e., DFT). The Dh-cp(DT) motif is predicted to be the ground state only at composition 22-12 at the empirical potential level, whereas it corresponds to the lowest energy structure at the DFT level for all the compositions (17-17 to 28-6) studied here. Segregation of Pd atoms to the surface of the cluster, at the composition $Pd_{17}Pt_{17}$, has been corroborated by our DFT calculations.

Gupta potential calculations on 98-atom Pd–Pt nanoalloys (using a combined GA and BHMC algorithms) indicate that the most stable structures adopt the Leary tetrahedron (LT) geometry, originally discovered as the global minimum structure for 98 atom Lennard–Jones clusters. Similar to our results for 34-atom Pd–Pt nanoalloys, Pd atoms are found to segregate to surface sites.

A combined EP-DFT system comparison approach allowed us to single out reasonable candidates for the lowest-energy structures at selected compositions for 38-atom clusters of four bimetallic systems (Pd–Pt, Ag–Pt, Pd–Au, and Ag–Au). For these four systems, the EP shows a preference for the TO arrangement, except for the Ag–Pt pair, for which an inc-Ih-Anti-Mackay structure is predicted as the GM. A peculiar Oh-Ih structure (originally found as a high-energy isomer for Ag–Au and pure Ag clusters [1]) corresponds to the putative GM for two of the systems studied in the present work, Pd–Pt and Ag–Pt, at composition 24-14. This confirms the usefulness of the system

L. O. Paz Borbón, *Computational Studies of Transition Metal Nanoalloys*,
Springer Theses, DOI: 10.1007/978-3-642-18012-5_9,
© Springer-Verlag Berlin Heidelberg 2011

comparison approach in searching for the lowest-energy structures of nanoalloy clusters [2, 3]. Moreover, DFT calculations predicts Au rather than Ag segregation at the surface in Ag–Au particles because of charge transfer effects (an effect not taken into account by the EP potential).

A detailed study of the structures and chemical ordering of 34-atom Pd–Pt nanoalloy clusters, as a function of composition and the strength of the heteroatomic interactions, shows a strong inter-dependence of chemical ordering and structure. Three main types of chemical ordering are observed: core–shell; spherical cap; and ball-and-cup (intermediate between the first two types).

Three parameterisations of the many-body Gupta empirical potential have been compared as regards the geometrical structures and homotops that they stabilise for Pd–Au nanoalloy clusters with up to 50 atoms, and for 98-atom clusters. The "average" parameters (where the Pd–Au parameters are obtained by averaging those for Pd–Pd and Au–Au interactions) are found to favour Pd$_{core}$Au$_{shell}$ segregation, as these maximise the number of the stronger Pd–Pd bonds and expose the most Au atoms (which have lower surface energies); while the fitted parameters, e.g. "exp-fit" and "DFT-fit", generally lead to more mixed Pd–Au nanocluster surfaces.

Finally, chemisorption of CO and H has been studied on metallic clusters (Pd, Pt and Au). It was found that the energy ordering of the structural motifs of Pd, Pt and Au clusters of size 38, is altered by CO and H chemisorption. CO chemisorption has been found to induce a "flattening" of the metal cluster PES for the three systems considered in this study, favoring non-crystalline (e.g., polyicosahedral) structures, whereas the effect of H adsorption is mostly connected with the release of surface stress.

These calculations on segregation effects as well as geometric properties of Pd–Pt, Ag–Pt, Pd–Au and Ag–Au (as well as chemisorption of CO and H on 38-atom metallic and bimetallic clusters) presented here, will hopefully be helpful in future catalysis studies (although the specific sizes studied in this work might not be the best candidates as catalysts). The combination of a genetic algorithm (GA), as an initial global optimisation technique, with the basin hopping atom-exchange (BHMC) approach, and systematic searching for high-symmetry shell structures, has proven to be a powerful technique for studying medium-size bimetallic clusters. DFT calculations are an invaluable tool for studying the energetics of nanolloys at the higher level of theory. In this sense, more first-principles calculations are needed (e.g. exploring different adsorption sites such as "bridge" or (111) hollow sites, as well as selective adsorption) for bimetallic systems in order to get a full understanding of catalytic properties as a function of nanoparticle composition and chemical ordering.

In future, studies of chemisorption and reaction of small molecules on supported nanoparticles will be undertaken, where molecule–nanoparticle and nanoparticle–substrate interactions will be considered at the same time. Such computational intensive calculations will be facilitated by massively parallel computer architectures based on increasingly powerful computer processors.

References

1. K. Michaelian, N. Rendón, I.L. Garzón, Phys. Rev. B **60**, 2000 (1999)
2. R. Ferrando, R.L. Johnston, A. Fortunelli, Phys. Chem. Chem. Phys. **10**, 640 (2008)
3. A. Rapallo, G. Rossi, R. Ferrando, A. Fortunelli, B.C. Curley, L.D. Lloyd, G.M. Tarbuck, R.L. Johnston, J. Chem. Phys. **122**, 194308 (2005)

Index

CPSIA information can be obtained
at www.ICGtesting.com
Printed in the USA
LVOW02*1152310116

473061LV00001B/94/P